Homesteading

Other Books by the Same Author:

The Wyeth People

Two Acre Eden

Gene Logsdon

Homesteading

How to Find New Independence on the Land

Rodale Press, Inc. Book Division
Emmaus, Pa.

International Standard Book Number 0-87857-068-3

Library of Congress Card Number 73-5159

First Printing—October, 1973
OB-214
Printed in the United States of America on recycled paper

Contents

Gene Logsdon—an organic homesteader

Chapter One

A New Kind of Farmer

You'd please me most just calling me a farmer and leaving it go at that. Most of my time is spent in agriculture one way or another. When not farming, I'm talking to farmers or writing about them. I like doing both.

I can't make any generalizations about the people who raise food for a living or to supplement their living, because I know too many of them: big farmers, little farmers, rich farmers, poor farmers, kind farmers, mean farmers, smart farmers, stupid farmers, family farmers, corporate farmers, commercial farmers, hobby farmers, pig farmers, dairy farmers, grain farmers, cotton farmers, rice farmers, vegetable farmers, fruit farmers, cowboys, ranchers, and migrant workers. I've learned to love them all, or at least admire them—I think they're one of the last tribes of independent thinking people left in the country. But I wouldn't want to press that generality too far either. I know some who haven't thought independently for years. Can't afford to.

There's another kind of farmer making his (and her) presence felt in the countryside now. I guess I'd call them the small-time farmers. Or the back-to-the-land city farmers. Or, what I prefer, the modern homesteaders. I've watched their numbers grow rapidly in the last couple of years, and I've listened with keen interest to what they've told me.

Who are these modern homesteaders? Talking to them, I get the feeling they come from two distinct groups. One group is coming out of cities and towns to live on small acreages where they hope to establish a certain level of subsistence while improving the quality of their lives. The other group is coming out of commercial agriculture—forced out sometimes—looking for jobs in some other industry but holding on to their country homes and a few acres. They will commute to work (as will most of the first group) while continuing to live the way country people like to live.

The two groups seem to have met each other while heading in opposite directions and decided to stop and get acquainted. The ex-urban society they form looks to me like the beginnings of a new Jeffersonian democracy. And we could use some of that these days.

I'm strongly in favor of modern homesteaders even if they do clutter up the countryside a little. I have to be in favor of them because I'm one of them—having belonged in turn to each of the two groups of which they're composed. My country life started at age five, when my father lost his business in the so-called "Great" Depression, and we rode the vanguard of *that* wave of "back to the landers" to the hopeful country.

I remember mostly from those early farming days how terribly hard my parents worked. I know why farmers buy machines they can't afford—you do it eventually out of bone weariness. But other than that, life had a certain unhurried peace to it, and I could observe that in my parents too. Father had time to take us arrowhead hunting in corn fields, or fishing in the pond. Mother often helped us build play dams in the creek across the road. Later, when we became hotshot commercial farmers with a modicum of financial success, there was no time for these things.

My earliest memory is a huge threshing machine powered by a steam engine. Farmers bring wheat bundles in from the fields on wagons pulled by horses—each man proud of his team. The wheat is tossed into the thresher's gaping maw, and the straw shoots out of the blower at the other end. Swarthy Ade Rall, our neighbor, red bandanna covering his mouth and nose from the dust and chaff, forms the straw into a perfectly symmetrical stack. Ade's still farming. He's probably the last man in America who can build a good strawstack.

A most vivid memory is of the time a team of horses ran away on me when I was nine years old; I remember it like yesterday. Anyone who has experienced the fear of a runaway never forgets. And I've preferred tractors ever since.

When I stand on the streets of Chicago or New York today, as I sometimes must, or when I am riding a jet high above the clouds, I think back to those days and am overwhelmed by an uncanny sense of loss. From a seat in a jet airliner, 1940, the way I lived it, was at least a hundred years ago. There is no other way to express the vast changes in my life when I look down at my hands fastening a seat belt and realize those same hands not so very long ago pulled the reins against the frenzy of a runaway team of horses. And those hands carried lanterns and coal oil lamps before they ever touched an electric light switch.

In the fifties, I went wandering, as all young men must wander and among the things I did to earn bread (literally) was to work on farms in five midwestern states. I learned a lot.

The last farmer I worked for was also the first one—my father. Then I was off again, in 1960, this time to look for a wife. Found a good 'un (as farmers used to say) and then I joined that group of small farmers coming *off* the land looking for steadier income.

I was able to catch on as a journalist, which is almost as risky as agriculture.

Just as quickly as I could rub two half-dollars together in my pocket, my wife and I bought our modern homestead. Being used to much larger acreages, I was somewhat disappointed that now in the area I had to live in, I could afford only two acres. But we soon found that you can raise an awful lot of food on two acres. All you have to do is know how. And want to do it.

And that is why I'm writing this book. The past events of my life seem now to have been leading up to it. We modern homesteaders want to know how to use the best of the old ways along with the best of the new ways to live as independently as possible on our land. Many of the old ways were my growing up ways and I think I know which are still practical; most of the new ways I have learned as I went along, and I think I know which ones to avoid.

What I know I write for all homesteaders, but particularly for those who are interested in organic farming and gardening methods. Many books have been written about how to raise food with all the chemical technological aids. Not many have been written for the *practical* food producer who does not want to use man-made, inorganic chemicals. Organic farming and gardening is certainly the more challenging way to raise food. Anybody who can read labels can do the job with chemicals. And working with organic homesteaders is often more interesting than with "conventional" gardeners or farmers. Organicists are so earnest, so enthusiastic, so adventuresome. They are willing to try new or unproved or even zany ideas.

Met a fellow last summer who had gone out and bought the

poorest farm in the county. The land was so bad commercial farmers weren't even bidding on it. The organicist was a rural mailman, and he had to use all the money he had so arduously saved to make a down payment. But he didn't mind—he was by far the happiest farmer I saw all last year, though he certainly wasn't making money on the farm.

But he was enjoying himself and experimenting with some novel ideas. For instance, he had concocted an insect spray for

Like most homesteaders, the Logsdons—Gene's wife Carol with children Jerry and Jenny—are enthusiastic and adventuresome. Their garden is productive.

apple trees made of laundry soap, milk, chalk, lime, rotenone, and ryania. "Ryania is a powder from the root of a plant that grows in South America," he told me, plainly fascinated by the knowledge. "It makes bugs sick to their stomachs, and so they won't eat very much." I suggested that his gardening was as crazy as mine, and he laughed and said, "But ain't it fun." The important fact is, his transparent apples and a couple of trees of an early fall red variety were absolutely unscathed by bugs.

Another "dares-to-be different" farmer grew organically an old open-pollinated variety of yellow flint corn to see if it had a better feeding quality for cattle and chickens than modern single-cross hybrids. He drew a lot of snickers from those who were convinced that hybrids are better than open I-I pollenated varieties, but sometimes adventuresomeness pays off. That year corn blight struck hard but did not affect the old non-hybrid varieties. He who laughs last. . . .

A beginning organicist I visited recently developed a weird but effective way to grow high quality melons. He lived far enough north (mid-Ohio) that it was necessary to get late-maturing melons started early indoors and then transplant them to the garden. He did not like this job. He figured out a way to plant them early in the garden.

He set old hay bales end to end across the garden, making two rows with about three feet between them. This space he filled with eight inches of manure and a covering of compost. In April he planted melon seeds in the compost, then put some old windows he had over the bales. In effect he made a hotbed of the whole garden row. Between the sun coming through the glass and the heating manure, the seeds sprouted and grew vigorously. They were pushing against the glass by the middle of May when the sash could safely be removed. In the meantime, the cucumber beetles had not gotten around to noticing the plants until the latter were in such excellent vigor, the bugs caused little ill effect.

By midsummer, the melon vines swarmed up and over the bales—which served as good mulch during dry periods that followed.

Organic farmers often seem to me more scientific than those scientists who make fun of them. Any idea not completely insane will get a test from organicists. They know that all the inventions we have today sounded crazy when first introduced. I'm sure that the most learned man of the fifteenth century would have howled with laughter if you tried to explain to him that someday you could look into a little box called a TV set and see people anywhere in the world.

Some organic gardeners play music to their plants "just to see" if it affects growth—and if you want to laugh at them, go right ahead. They don't care. Others experiment with the influence of electricity and magnetism on plants, "just to see" what happens. Still others will "naively" try any mineral soil conditioner that comes on the market. But why not? Soil science is not a closed book, forever ended. And while the know-it-alls are laughing, some organic gardener, probably one with little previous experience, will discover something important—something the scientists might have found had they not been laughing so hard.

It amuses me that scientists at the Department of Agriculture's research headquarters in Beltsville, Maryland, have finally discovered that beer can control slugs (a practice which incidentally doesn't work very well for me). Organicists have been using beer to kill slugs for years, and the scientists have finally decided to take them seriously.

Other organicists I know experiment with animal nutrition, too. One read how New Zealand dairymen "graze" their cows on mangel and turnip roots, getting good milk production with less grain feeding. He applied the idea to his rabbits. He planted extra rows of carrots, covered them in winter against freezing, and fed the vegetable to his nursing does. The young nursing rabbits grew better than ever, and he was able to save half his bill on high-priced pellets.

Incidentally, if you want to try feeding mangels to cows, you can get seed from R. H. Shumway Seedsman, Rockford, Illinois 61101. Their catalog says that Colossal Long Red mangels yield thirty-five tons to the acre—a lot of feed. They can be fed to all livestock. In times of high grain prices, such as we saw in 1972–73, feeding roots may not be such an old-fashioned idea.

I visited another organic farmer who decided years ago to replace all chemical fertilizer on his place with a mineral soil conditioner and see what would happen. What happened is that his soybean crop the year I visited him yielded fifty-one bushels to the acre, which is very good. I waded through the beans when they were taller than my waist. I have to admit I still don't quite believe what I saw, but I have a picture to prove it wasn't a dream.

When I tell certain commercial farmer friends of mine about this imaginative experimentation of organic farmers, they are apt to snort and call it something else, like "a great deal of foolishness." Sometimes I argue, sometimes I don't. But one thing I've noticed which is not foolishness no matter which side of the fence you are on. More and more it is the modern homesteaders, not the typical commercial family farmers, who tell me how much they enjoy farm *life*, how much they revere the rural virtues of hard work, independence, self-reliance, thrift, and love of land. It is the new homesteaders who tell me most often about how wonderful it is to raise a family in the country; about what an opportunity for work and play a farm provides for parents and

The farm life draws the organic homesteader to an independent, sometimes hard, life close to the land.

growing children who need to establish a sense of place and a sense of responsibility; about what a satisfying feeling it is to raise one's own food; or to look across a field well-cultivated; or to be able to spend a spring day working with a son. The commercial farmer seems to have to work so hard, he no longer enjoys the land as much as he used to.

I am both encouraged and dismayed. Encouraged because I see the possibility of a new kind of family "farmer" emerging to carry on the spirit of Jeffersonian democracy which economics has beat out of so many commercial farmers. Dismayed because I have so much sympathy with the commercial farmers like one who told me recently: "Why don't you write an article for *farmers* on how to get more time with their family. It just seems like I have to work all the time." This very fellow is now leaving commercial agriculture; he could not make enough money at it.

I have been in both groups: "city" farmer and country farmer. I wish they could get to know each other. Perhaps through the pages of this book they can meet and understand they share common goals.

Chapter Two

What a Real Homestead Is Like and Where to Find One

When the city man dreams of living in the country on his own homestead, he romanticizes. He thinks how nice it would be to have his own milk, butter and cheese, avoiding in his mind the fact that a cow must be milked twice a day every day. He envisions his livestock producing all the manure he's always wanted for his garden. He forgets a pile of manure may stink if not handled correctly. Raising pigs for his own pork can mean raising his own flies, too. His own source of water means he has to tend the pump.

When a commercial farmer dreams about quitting the rat race which now effects his farming (even to a greater degree then some city businesses), he romanticizes, too, but in another way. He figures he will get a job in town, at least part-time, to satisfy the living needs of his family. He will operate his small acreage he has retained *for fun*, and any profits will be gravy money.

Drawing on his experiences he envisions the "perfect" small farm. Forty acres in good farm country, neither too far south nor too far north, and where rainfall is adequate for crops without irrigation. The soil is a deep, fertile, limestone clayey silt loam. A never-failing spring originates on the property, trickling away into an unpolluted stream. Eight acres are in woodland, left to grow naturally. Ten acres of good, improved pasture lie along the stream. The rest of the land is well-fenced crop fields, gardens and buildings. The land rolls gently toward the creek for good soil and air drainage, but not so steep as to cause erosion problems. There's a good well on the place. Also a solid big barn, well-built eighty years ago. The house is modern, in good repair, but modest so as not to send the price of the property out of reach. The chief, and pratically only, industry of the area is a viable kind of commercial agriculture, preferably the stable family farm enterprise found in dairy farm country or the diversified grain livestock farms of the midwest.

The city and the farm dream of an organic homestead fall short of the reality, and I say that not to discourage you, but

Everyone dreams of the "perfect" small farm, with good pastures, fertile crop fields, woodland, never-failing stream, and solid buildings.

to warn you. Life in the country can be beautiful; perfection doesn't exist.

But there are real homesteads that come close. Here's one, for example, whose progress over the last six years I have watched with amazement.

Joe and Sarah Smith (not their real names) were both born and raised on farms, which put them a leg up for a homesteading venture. Joe's father had other sons and only a small farm, so Joe went to work in another business. He sold commercial livestock feed and farm machinery, both of which jobs further broadened his knowledge in a way that would be helpful in modern homesteading.

Soon after he and Sarah married, he bought an old frame farm house, a barn, and other outbuildings plus ten acres to try out ideas in organic food raising, a subject that had begun to interest him. "Gradually" became the key word in their conversion to organic country living. They gradually renovated the house while living in it. Some buildings were torn down and others fixed gradually, while Joe continued his full-time selling job away from the farm. Gradually he and Sarah expanded vegetable, fruit, and livestock enterprises until they (five young children) were raising almost all the food they needed, with some to spare. Listing the variety of foods the Smiths produce sounds like a role call of the Horn of Plenty: Every vegetable that will grow in Indiana, plus a few that aren't supposed to grow there; four kinds of berries, grapes, melons, apples, peaches, pears, and plums; chickens for meat and eggs; rabbits and turkeys; several litters of pigs a year—enough to sell to cover the cost of grain they eat; one or two calves annually, fed out and butchered for beef; bees for honey. And as if that were not enough, Joe built a small pond and stocked it. Ducks swim on the water, bass swim in the water (not to mention the family and two dogs).

Because he lives close to dairy farms, Joe has been able to

get an almost unlimited supply of manure for the hauling. Thus he has been able to indulge his organic interests to the utmost. He puts manure a foot thick on garden plots and tills it in. When he started his orchard with fifty young trees, he mulched all of them with manure and credits that with saving most of them in a very dry first year. He also has good luck growing potatoes under straw mulch rather than underground.

Last year he started raising fishworms to seed his garden, to compost his rabbit manure, and hopefully to sell to fishermen. He has also built a small greenhouse to grow winter lettuce in. He is so enthusiastic about trying new organic ideas, I'm afraid if *Organic Gardening and Farming* magazine ran an article on growing cabbages upside down, he'd try it.

The key to the success of this operation is, of course, work. You have to like to do it. Sarah especially seems to enjoy, as few women would, the challenge of "putting up" all that food. Hardly a week goes by, summer and fall, that she is not freezing, canning, or jellying something. Her jellies have intriguing names like Crabapple Burst and Eve's Temptation.

Joe calls the wine he makes Organic Eye-Opener. He makes cider, too, takes the honey from the hive, smokes hams and sausage, dreams up "more fool ideas" in his words.

It was one more fool idea that prompted him this year to move on to the third phase of his plans: he bought a small country general store a couple of miles away and quit the selling job that had kept him on the road travelling so much of the time. He and Sarah hope eventually to add an organic specialty section to their store, through which they can sell the produce from their homestead.

With all the work, you might think the family has no time for play. Not so. Vacations each year are usually spent camping and/or hunting, and often turn into food-gathering expeditions. Besides wild game, the Smiths are particularly fond of morel mushrooms, for which they will drive considerable distances to hunt. They dry the morels they don't eat right away, then freeze them for year round use. The children gather wild berries in season; the oldest boy traps muskrats for spending money and keeps the larder full of squirrel during the season.

No miracles involved here. The homestead cost the Smiths about $10,000, and is now worth three times that amount. They've sunk about $10,000 more cash dollars into the place since buying it. With the store now, they have some debt, but it's healthy debt mostly on income producing or appreciating property. "Right now, I am very happy," says Joe. "No more travelling, no more nights away from home. We aren't going to get rich, but we're going to enjoy life."

What is most amazing to me is that this couple is not as unusual as you might think. They are unusual—as are all homesteaders—but not atypical. I know two other families whose pattern of life is almost identical to the Smiths. And I haven't really tried to find such people. There are thousands of them—and you can be one, too.

Where to look for your place? Any area. It would be naive for me to suggest that one section of the country offers better homestead properties than another. For my money, Lancaster County, Pennsylvania, is ideal, but urbanization is rapidly closing in there. The bluegrass region of Kentucky is another almost perfect place as to weather, soil, and living conditions. But the rich know that, too, and you almost must be wealthy enough to own a race horse farm to buy land there. Southern Wisconsin and most parts of Iowa are also excellent places to start a modern homestead if you don't mind cold winters and if you can find a farmer willing to sell good acreage. That is sometimes a big if. North Carolina, surprisingly, has more farms than any other state except Iowa, which means, of course, that the farms are small. It's a good place and a good climate to start modern homesteading.

Actually, it's ridiculous to mention specific areas for good organic homesteading. Suitable places are everywhere, in every state. Even in localities generally called poor farm land, you can find spots that are quite adequate. For instance, the Appalachians are not especially well-suited for general farm-

Practically every state offers some good homesteading spots. Find a place you like and look; rent while you look if you must.

ing, but the mountains are broken up by numerous hollows where small plots of rich soil do exist—and good water, too, unless the strip miners have gotten there before you.

Then too, even poor land can sometimes be made fertile, especially by organic methods. All it takes is more time and a little more expense.

The only rule you can follow is to go to the area you want to live in, and *hunt.* Better to rent there awhile before you buy. Follow the usual advice: Work through an attorney, trusted realtor or natives who won't lie to you; be sure you get clear title.

People moving into a new area are always counselled to look for good schools if they have children. I have doubts about that kind of advice. What, pray tell me, is a good school? Some of the best citizens I know came out of schools that by today's educational standards were below par. There are an amazing number of successful men today who never completed high school at all. You don't hear about them because that's bad publicity for our educational establishment. Personally, after going to and teaching in schools longer than I care to recall, I think they're all mediocre. A child gets educated in spite of them.

You are also advised not to move into poverty areas or rural ghettoes. Well, I *do* know whole families of thieves who live in rundown rural houses, but I also know whole families of thieves who live in very fine suburban houses. Ditto for drunkards, prostitutes and liars. So how do you protect yourself from reality? Packertown, Indiana, is one of the ugliest rural ghettoes in the country. It is full of wonderful people and beautiful children. The percentage of undesirables there, is about what it is in the rich suburbs of Los Angeles.

I think of forty acres as the ideal, but that is a negotiable figure. Five acres serves very well indeed. Or two. For purposes of food, the acreage you need is directly related to the livestock you intend to keep. If you want to produce your own beef, pork, or mutton, you should have at least five acres. Chickens and rabbits require only two acres, and you can get by on less. You can raise enough fruit and vegetables for a family of four on half an acre.

Remember too that an acre of rich land is worth two of poor. In the south, you can grow more than one crop a year on the same land. In Alaska, you will get one, if you're lucky, but you can more easily add wild meat and fish to your larder there than you might in Alabama.

When people think of escaping cities for country life, they often visualize far away places where population is sparse. I don't recommend such a move, especially if you are used to an urban neighborhood. It's good to have a few neighbors within sight. Isolation is a mixed blessing, if a blessing at all.

Some of the best homesteading sites I know lie in the so-called megapolis between Boston and Washington. You might have to put up with crowds on the highways and beaches, but properly landscaped (and I don't mean fancy or expensive) outer suburban sites along the east coast preserve for their owners more privacy than a 200-acre farm in Illinois does. The same can be true of any suburban area where lot sizes run larger than an acre.

Younger people might want to venture far out into the country where land is cheaper to find their Eden, but older folks who need doctors and other services close at hand can find excellent, reasonably priced acreages with house and barn, right within the incorporated limits of many small towns and villages. This is very true of Pennsylvania, Ohio, Indiana, and south-central Minnesota, so I presume it is the case everywhere. But you have to hunt for these places—an exciting adventure in itself.

About eighty thousand farms and farm tracts are sold every year in the United States, and the best place to find out about them is through the local newspapers. Many magazines also carry farm-for-sale classified ads. Very little public land suitable for farming is left available for free homesteading. But if you do find some, you can farm it if the Bureau of Land Management, Department of the Interior (in Washington, D.C. where you should write for information) has classified it as suitable for farming. The trouble is, the Bureau judges such land on the basis of its potential for *commercial* farming, which isn't what the organic homesteader has in mind usually. Some public tracts do come up for sale, but most are in the semi-arid parts of western states. A hardy couple might make it there, but I don't recommend it. The only practical

Excellent homestead sites complete with house and barn, can sometimes be found in villages or small towns.

reason for trying this type of homesteading would be a lack of money to buy anything else. But if you lack money altogether, organic homesteading isn't a very good idea, as I will explain later. Some land in Alaska can still be homesteaded. Some people try it. Some come back in a year, too.

While I'm on the subject of psychological advantages and disadvantages, take a tip. DON'T begin your adventure in homesteading in the late fall or winter. Move out in the spring or summer. You are going to learn a lot of things the hard way the first few months you live on the land, and bleak winter weather won't help. At least if you move in the summer, you won't freeze if the furnace breaks down.

I want to repeat. The only proper way to get your place is to go where you want to live and hunt. Spend as much time as you can. Don't be hasty. I once almost bought an idyllic twenty acres out in the country far from city pollution. On my fourth visit to the isolated tract, the wind was blowing from the south, and that's how I learned about the rendering plant below a hill in that direction. Compared to the aroma of a rendering plant, I'd sooner endure the refinery smells in south Philadelphia.

Your choice of a place of your own, in the final analysis, depends upon how much money you can or want to spend, and somewhat on the taxes you will have to pay on it while you live there. That's what the next chapter is about. Money.

Chapter Three

Money and Other Gross Considerations

If you already are established on your organic homestead, you've followed the advice in this chapter and succeeded. Or you have not followed it and failed. So you may not need to read it. What is said here is for the person just planning to move to the country.

I assume you are not rich. If you are, you can buy any darn thing you want and outlive the consequences. My advice to the rich (which is never taken because no one ever believes he is really well-off enough to be "rich") is to salt away $300,000 in an insured savings account, buy a modest home and a few acres in the country and live in frugal contentment on the interest from your savings plus what money you can make potting around in a greenhouse, roadside stand or craft enterprise. Let your place support your basic needs by the power of your own hands, and spend your spare time doing good in your own neighborhood. You can have as full and meaningful a life as you desire—and you'll live ten years longer.

Anyway, I assume you do not have $300,000 or even $10,000 salted away. You have a demonstrated ability to earn a living wage and the discipline to hold a job and budget some savings.

(Without these abilities, you won't make it on an organic homestead anyway.) Beyond that, you should have at least $3,000 in savings, plus some equity in your present home. If you have no home now, you ought to have up to $10,000 in savings or the capability of borrowing that much from a kind father or father-in-law or someone else willing to hold a second mortgage.

Buying country property is usually (I must keep using that term "usually" because there are exceptions to everything) not the same as buying a house in a subdivision. Country tracts usually require a down payment of one third the purchase price, while in a subdivision you need only one-tenth. Usually.

According to the 1972 statistics from the Department of Agriculture, farm real estate value *averaged* $205 per acre, an increase of five percent since 1969. Land increased nine percent in the Northeast since then, only two percent in the North Plains, with the rest of the country somewhere in between. This doesn't tell you much. The price you pay for a house and small acreage in the country is more directly based on the quality of the house than the land. Buying a fairly decent house with from two to ten acres should run something like from $12,000 in Oklahoma to $40,000 in eastern Pennsylvania. That doesn't tell you a whole lot either. A specific example: You may be able to buy a central Illinois farm of two hundred acres, buildings included, for $400 an acre. But if you buy just the house and five acres, you'll pay more than the $400 an acre rate. You'll pay somewhere between $15,000 and $30,000 depending upon the condition of the house and the owner's eagerness to sell. If it's a real sumptuous place, you'll pay more. If it's falling down, you'll pay less.

You'll pay more if the place is close to a city and feeling the pressure of suburban development. You'll pay less if the homestead is out in the so-called sticks.

An abandoned farmstead with dilapidated house, barn, and outbuildings can sound cheap and inviting, but fixing it up can be a ruinous burden.

If you buy "bare land"—land without a building on it—and you buy, say five acres, expect to pay $1,000 an acre or more, if the going farm rate is half of that. It is much easier to buy small acreage *with* a house on it from a farmer than it is to buy bare land. An old house is usually a burden to a farmer. The rent is hardly worth it for him, and he has to pay taxes on buildings, even empty buildings.

You won't save much money buying a rundown place for under $10,000 either, because you will spend another $15,000 for repairs. You can get by for less, of course, if you make the repairs yourself (assuming that you can), but that is usually a poor idea. When you move, you will most likely go right into another forty hour a week job (I hope) which means that your wife will go out of her mind contending with leaky roofs, broken water pipes, crumbling walls, and sagging doors until you get around to the repairs or get her around to them. Either way, you are working up to divorce, desertion, family feuds, or a move back to the city.

A "for instance." As I write this, I just happen to know of a farmhouse, barn, and two acres for sale right now in a good location in Ohio for $6,000. Sounds great, doesn't it? The house is totally decrepit, and the barn is ready to go down with the next strong wind. That place will take $18,000 to put into a shape you'd be proud to live with.

I like to use the ballpark figure of $25,000 as the cost of buying just an adequate country place. Some people know that's too low for their area, others will think it too high. But it's still a pretty good jumping off figure.

Let's say you bought your present home ten years ago for $15,000. You can get, if inflation has hit your area like it has mine and if you've kept the place up, at least $20,000 for it today, after deducting brokerage fees and so on. Let's say you've paid off $4,000 on the place which gives you a total equity of about $9,000, more than the one third downpayment on your $25,000 place in the country. Your banker should be glad to go along with you as soon as he knows you have a job in your new area.

Be sure you have an attorney to help you make the switch so you can take advantage of all tax breaks available. For example, you will not have to pay income tax on the "profit" you make selling your present home, if you spend that increase in capital assets on your new place. In other words, if you can get $25,000 for your present home, you might as well buy a place priced at that amount. If you buy a cheaper one, you'll pay taxes on the difference.

If you can't swing a deal through a bank, don't despair immediately. Get in touch with your nearest Farmers Home Administration (FHA) office. Get to know the representative. Lay your cards on the table. You may qualify for a low interest loan.

If you can't swing any deal, common sense should dictate that you are not ready yet for a move to the country. Do not move in poverty on the strength of dreams. It won't work.

If you haven't owned a home before, you will cringe when you figure out how much interest you'll be paying on your mortgage. Remember that this interest is deductible for income tax purposes, which in effect, reduces an interest rate of, say, seven percent to something closer to five percent.

Another thing to remember. Don't be afraid to buy land if you can swing it. Historically, land prices go up. Sometimes fast, sometimes slowly, but always up. Buying land, if it doesn't leave you without cash for living expenses, is just as good as putting money in savings and usually is better because investment in land is a better hedge against inflation. Unlike cars, land rarely depreciates. Unlike houses, you don't have to spend a lot of money every year to keep land from falling apart. I tell you this because you may be faced with a situation where you can buy a home and one acre with the offer of "more land if desired." Take as much as you can prudently afford. Sometimes you can get additional acres fairly cheap this way, if you drive a good bargain. That extra land, deemed cheap by other standards, is worth its weight in gold for the organic homesteader and is always a safe investment.

Property taxes can become a crucial burden for retired people on fixed income. For this reason I will, in a later chapter, show you how you can make your small acreage pay you enough money to offset rising property taxes when you are retired. A younger owner should not be too upset by rising property taxes, because it means the value of his property is rising, too. However, property taxes are becoming unbearable in many areas because of the school taxes, which are the bulk of the property tax.

A word of caution. When you find a place advertised for sale, there's almost always a reason other than "a tragedy in the family." Otherwise really good buys are snapped up by area residents who know what the score is. *Always* remember that. When you buy, you will have to pay a little more than the place is worth, or you will, by and by, find something wrong with the place. Nine times out of ten, you can correct what is wrong. But be prepared. If the place was a good deal at the offered price, it would never have to be advertised in the paper.

Of course, a piece of property is often sold at auction as a way of settling an estate. You may get a good buy this way, but bidding for land (or anything) at auction can be taking your financial life in your hands. Take it from one who knows: auctions are very often rigged—very legally of course. The owner and/or auctioneer arrive ahead of time at a base price below which they do not intend to sell the property. Sometimes this is openly stated. Other times, a stand-in will simply join the bidding until the price is right. Sometimes the owner ends up holding the bag, but not very often. Know beforehand what you can afford to pay, and don't let the emotion of the auction change your mind. If you have informed the owner or auctioneer that you are interested in the property, or if they know that anyway, it is a very good idea to let them know how high you will bid. You may think this is tipping your hand, but if your top price is a fair, reasonable one, you will be better off than being secretive, hoping to get a bargain. If they know your limit, they will not be tempted to run the bid over your head (unless they think you're bluffing), at least not without another bona fide bidder who will go higher. In the latter case, you probably wouldn't get the high bid anyway. Just remember that bargains are scarce as hen's teeth at land auctions. The best you can hope for is a fair purchase price. More than likely, as in bidding for antiques at auction, you'll not get any bargains.

One way you might, though, is at auctions where a farm is sold in sections. Let's say an eighty acre farm is being gavelled off in two twenty-acre plots and one fifteen-acre tract, all suitable for subdividing by a builder. But a fourth plot, consisting of the other five acres, might lay in an irregular shape bounded by a creek or other natural or man-made obstruction that makes the acreage expensive to develop. You might get a comparatively excellent price, bidding on that chunk.

When looking at a country place, one of the first things to check is the well. Is it a good one? If not, is it likely that a good well can be dug? If you have to drill a new well, find out what it will cost from a well-digger and budget for that. Some places you can hit good water forty feet down; in others you'll go four hundred or more, and the dollar bills start adding up. But you have to have a dependable source of pure water. Remember too, that there are good organic homesites supplied by city water.

Your little homestead will require a few implements right away. Normally, you would figure on a riding lawn mower with attachments (about $1,200), a rotary tiller (about $150 to $350), a grinder-shredder (another hundred or more), a snow plow or blower ($50 to $300), and some kind of cart or wagon to haul things in. In other words, you're talking about something in excess of $2,000. It may sound like heresy, but my advice, especially if you have two acres or more, is *not* to buy these implements. Instead go to a farm implement dealer, and price used farm tractors in the eighteen to twenty-five horsepower range. Often you can get one, with attachments to do the work of all the above mentioned tools, for $1,000 or less. And with such a tractor, you are equipped to handle fifty acres, if you want to.

I don't want to sound like I think one brand of tractor is better than others, but in my experience the old Ford-Ferguson tractors made in the fifties, Models 8N and 9N, fit the bill very well. The Ford and Ferguson people were the first to perfect hydraulic lifts for their tractors, and this is why these models, which run about twenty horsepower, are ideal for the small homestead. They handle a two-bottom plow, a small disk, rotary mower, cutter bar, 'dozer blade, and front end scoop hydraulically. This means you can raise and lower the implements at will, which is extremely handy for cultivating garden plots and mowing lawns that might otherwise be too small for this size tractor. You can sometimes buy an 8N or 9N with the above mentioned tools for about $900, sometimes less than that at a farm sale. You may have to do some overhaul work, but these tractors have proved to be quite durable. If you do buy an old tractor that has been sitting out in the weather awhile without good care, make sure the block isn't cracked. If you don't know what that means, take someone with you who does.

The two-bottom plow and the disk that comes with the tractor will work up your garden much better than any rotary tiller, and almost as handily. The tractor-driven rotary mower will work well as a mulcher-grinder-shredder. Just drop it down over a pile of leaves or whatever, and it will do the job. The 'dozer blade will clear lanes of snow. Out in the country, especially if your place has a fairly long lane to the road, a 'dozer blade is almost a necessity. The little snowblowers will do the job, but in a far greater length of time.

Incidentally, if you do have a lane, very often you'll have to put down some gravel before the spring thaw. Budget a couple hundred dollars for this purpose.

More recent models of Ford tractors of the size we're talking about are of course, even better, but you will pay more. Most farm machinery companies now have similar models which you can buy used, also.

If you have the money, better to buy a new twenty horsepower farm tractor and attachments, than smaller lawn and garden equipment. Such farm tractors will last years and years longer and save you lots of money.

If your homestead is under two acres, take a look at the new electric, battery-driven lawn and garden tractors before you buy the noisy, carbon monoxide belching gasoline jobs. General Electric's ElecTrak has attachments for most outdoor chores, performs as well as gasoline models, and is delightfully quiet. If all suburbanites used them, Saturday in the suburbs would become endurable again. All you have to do is plug in the battery to ordinary house current overnight, and it recharges. Cost of battery-driven tractors is comparable to gasoline models of the same horsepower.

A good used farm tractor is a better buy than a new garden tractor for the homesteader. For the price of a garden tractor, you'll probably be able to buy a more rugged and powerful used farm tractor with an array of implements. Try to get a shop manual if you can and do your own maintenance work.

Living in the country, unfortunately, makes you more dependent on the automobile, especially if you have a job more than two miles away. In fact, you will probably need two cars if you or your spouse remains at home while you are working in town. If you are a two-car family, BE SURE THAT YOUR SECOND CAR IS A PICKUP TRUCK.

For the organic homestead, no tool is more necessary than a pickup. You'll need it to haul mulches, manures, and other natural fertilizers. And a hundred other uses. It doesn't have to be a new one. Buy a used one, even if you have to put a rebuilt motor in it. Economists won't always agree that buy-

You can get by without a pickup, but if yours is a two-car family, why try? A pickup can do everything from plowing the snow to towing a livestock trailer to the slaughterhouse. The sweptside model shown is most popular amongst farmers, but the traditional stepside model offers a bed with flat sides to stack against. The options and accessories available can run the cost of a truck into luxury-car range, so shop with an eye to practicality.

ing used equipment is the most efficient use of money, but that's because economists haven't had to scrounge. If you have to borrow money to buy new, don't. Buy used. The more money you owe, the less chance you'll ever gain independence on your homestead.

Do not buy any livestock until you are ready and know what you are doing. More on that later.

Budget money for your first year's plantings. Your vegetable seeds will cost you about $30, if you buy good seed. You should put in about twenty-five fruit trees the first spring and figure on a $100 for that. Another $30 for bush fruits and strawberries just for your family. After you get some experience, you may want to grow food for market and crops for cattle, in which cases see appropriate later chapters.

Buy a freezer, one about fourteen cubic feet in size, to begin to take excess garden produce in June. If you intend to raise your own meat, and you should, buy a second freezer, same size, when you're ready for it. I think it is better to have two smaller freezers rather than one large one, but that's a matter of opinion.

I've read studies that claim a freezer doesn't pay for itself, but they were made before food prices started skyrocketing. A freezer is like any other piece of equipment. If you use it to capacity, it jolly well does pay for itself. If you only half fill it, then don't use the food you store in it (it happens, believe it or not), then of course it doesn't pay.

We'll talk about other low-cost storage methods later on, but from a practical point of view, a freezer is almost a must on an organic homestead.

Buy a hammock. If you can't find time to lie in it and watch the birds on a hot afternoon, somehow your organic homestead ain't making it.

Chapter Four

Soil—The Source of Life

The secret of successful food-raising lies in the soil. If the ground you grow your vegetables, fruits, and grains on is properly conditioned and cared for, your work is half completed. You don't grow the plants, really; nature does. Take care of your soil and the plants will take care of you.

It is difficult to spot good land simply by looking at it, unless you are a pro. Color won't help much, because red, brown, yellow, and black soil colors can denote either good land or poor. Sandy soil has its problems, but properly handled, it can grow good crops. Clay soils have their problems, too, but good organic methods can overcome them. The experienced can judge a good brown silt loam (the best all-around kind of soil) by its feel. Also, by examining a vertical creek bank or ditch, or simply by digging a hole in the ground, you can ascertain how deep the topsoil is.

A surer way to judge soil quality is by what is growing on it naturally, if you know your botany. Some plants grow only where soil is wet, like swamp oak. Cedars grown, or like to grow, on acid soil. Rather, more correctly, cedars will grow on worn out, acidy soils where hardwoods won't. When, for instance, you see fields in eastern Pennsylvania dotted with cedar trees, you can bank on the fact, that if you clear and farm that ground, you will have to lime and fertilize it heavily to get good crops of corn or legumes.

The pioneers said that land that supported large black walnut trees was always good fertile ground. In my experience, they were correct. But generally speaking, land where anything grows extra big and lush, even weeds—particularly weeds like bull thistles, burdock, ragweed, which like well-drained soil—indicate land potentially good for farming. Stunted plants, of course, indicate the opposite. However, certain cultivated crops have their own rules. The best wine grapes, I'm told, grow on naturally poor soil. Blueberries do well on sandy, acid soils unsuitable for grain crops.

The surest test for potentially good soil is to look at gardens and farms already established in your area of interest. Locales

An experienced hand can judge a soil by its feel. You can learn a lot about the texture, structure, and consistency of a soil just by rubbing between the thumb and fingers. Sand particles are gritty. Silt has a floury or talcum-powder feel when dry and is only moderately plastic when moist. Clayey material is coarse when dry and very plastic and sticky when wet.

where neat and substantial farmsteads are the rule usually mean excellent soil. Or if a neighbor with the same kind of soil as yours is already growing an abundant garden, you should be able to do likewise. Other tip-offs to good and poor soil: when dock, pigweed, lambsquarter and purslane *thrive* in a garden and ragweeds grow rank in the fence row, the soil has good organic content, is fairly well-drained, at least, and is fertile. If sorrels and fennel (mayweed, dog fennel, chamomile) make up the majority of your weeds, your soil is likely acid, low in humus, and lacking in fertility. Buttercups, ferns, trumpetweed (Joe Pye weed), iron weed, loosestrife, and Creeping Charlie mean the soil is probably fertile but needs drainage. Mosses, lichens, and poverty grass indicate poor

Give primary consideration to the soil when looking for a homestead, for its quality can determine the success or failure of your homestead. The existing vegetation can give you clues about the soil quality. Relatively bare land with only patches of scrawny weeds is a good indication of poor soil. A better homestead site would be in an area blessed with naturally fertile soil, indicated by the heavy cover of weeds and grasses.

soil. Thank my grandfather for that wisdom, not me. I don't know a couple of those weeds, and I've seen exceptions to what he says of a couple more. But one did not argue with my grandfather. Not on farming. He was quite successful at it.

In any event, land not hopelessly alkaline or acid, swampy or steep can be brought into fertility. It entails three basic steps:

1. Drainage. An old farm adage says, "Good farming is thrown away on wet land." That's gospel.

2. Erosion control. For soil conservation purposes, land is classified by its slope—level, rolling, very steep. Each type involves specific soil practices to prevent erosion. Also wind erosion is more severe on some soil types than on others; specific cultivation practices can prevent it.

3. Fertilization. Almost all soils lack certain nutrients, minerals, or organic matter—or have an imbalance of one nutrient over others. Or soil acidity may be preventing nutrients from becoming available for plant use. You must provide a *balanced* organic fertility and lime. Nature doesn't always know best.

To those three steps, one might add that the gardener or farmer must provide moisture for the plants, or at least preserve moisture as much as he can for periods of drouth. Mulch does that; fine cultivation does that; a good organic tilth to the soil does that. Irrigation remedies situations or areas lacking sufficient rain.

One of the first farms I worked on contained almost every soil condition and problem you will face—whether on a small farm or in a garden. How these various problems were dealt with makes a good study for the homesteader. This farm stood partly on rolling bluffs above a river valley. On top of the bluffs there were both steep and gently rolling clay hills. Interspersed among these hills were swampy potholes into which run-off water from the hills drained. The farm continued on down the side of the steep bluffs into the river valley. At the foot of the bluffs were fields of light sandy soil, then a stretch of spring-fed swamps. Beyond the swamps lay a low ridge of black sandy soil which fell away to black loam "river bottom" land next to the river.

For each of these soil situations, we developed specific practices, endeavoring to work with nature rather than against the tough old lady.

When I first worked at this farm, the clay hills were in a rotation of alfalfa, corn, and oats. In years when corn occupied the hillsides, a rain would carry tons of soil down the slopes. I was appalled, after one had summer rain, to see corn seedlings at the base of the hills completely buried in six inches of mud.

On the gentler hills, we switched to strip-farming—planting the corn, hay, and small grain crops in alternate strips parallel to the slope and following the contour of the hills.

"Strip" farming and terracing decrease erosion dramatically. The Department of Agriculture's soil conservation agent in your county will be glad to advise you in detail on these practices. If you have purchased an old farm, you will find the Soil Conservation Service most helpful in planning erosion control. Remember that soil erosion is still the number one pollutant in this country.

We took the steepest hills out of cultivation completely, using them for pasture, for plantings of Christmas trees, and for an orchard.

First, the pastures. We manured and limed the hills, planted brome grass and alfalfa, added rock phosphate. To bring fertility back fast, we took no hay crop off and allowed no grazing for two years. The grass was cut with a flail chopper and allowed to rot back into the soil as organic matter. Often a farmer thinks that if he plants a legume or grass in rotation with field crops, he automatically renews the soil.

Strip-cropping involves planting crops in alternating strips parallel to the slope and following the contour of the hills. Its purpose is to prevent soil erosion.

But if he removes all the hay or grazes the pasture hard, he is depleting the soil just as badly as a corn crop would, or worse, because with corn, often only the grain is removed, while the stalks and leaves are worked back into the soil. When a farmer removes all the grass, he must fertilize heavily just to stay even. Little organic matter gets into the soil (except from roots), and little real improvement comes to worn hillsides.

The dilemma comes because the ordinary working farmer thinks he cannot afford to leave that land "idle" until he has brought it back to good natural fertility. And the truth is that he often cannot and still meet his yearly financial obligations. The land is farmed to death, sold, resold, until finally (or hopefully) a farmer who has financial depth invests in future fertility rather than in mining the soil to break even. This is one reason why I think it is so necessary for the organic homesteader to make his living, or a substantial part of it, from an outside job. That way he is not pressured into soil mining by his mortgage.

Once we had a good pasture established on the badly eroded hills, we could allow cattle to graze them, being careful not to overgraze. We could, in effect, "break even" at a profit now, because we were working this land *at a higher level of fertility.* Without chemical fertilizers, we could harvest by grazing about three tons of grass per acre and allow another ton and a half or more to mulch back into the soil for fertility. Formerly, the hills produced only a *total* of three and a half tons, *all* of which was taken off in hay—which meant that yield steadily decreased with fertility.

One of the steeper hills became an orchard. Fruit trees do well on a hillside because of air drainage. Air flows downhill much like water does. The effect on air around fruit trees is to raise the temperature slightly—just enough sometimes to prevent tender fruit buds from freezing in winter. Also, in spring this air movement can prevent frost from killing blossoms.

We chose a north-facing slope for the orchard because the sun in winter and early spring does not strike a northern exposure as strongly as it does a southern slope. On a southerly slope, winter sun can heat up a fruit tree until its trunk temperature is ninety degrees even though the true air temperature is in the thirties or forties. Sap can begin to rise then, and if night time temperature drops quickly to zero or close to zero, as can often happen, fruit buds, which may have

Hills and fruit trees are natural companions. During still nights, the cold air drains from high lands into depressions and valleys. A difference of ten feet in elevation can make a five- to ten-degree minimum-temperature difference. In some seasons, this could make the difference between a full crop and crop failure.

begun to swell, are likely to freeze. On a northern slope, trees are naturally protected from this danger to a great extent. More importantly, because the trees don't warm up so fast on a north slope, blossoming in spring is held back until after the worst danger of frost has passed.

With the hills returned to good natural farming practices, we turned our attention to the swampy "potholes" between the hills. One pothole we allowed to continue as an upland swamp. Wildlife loved it, and we figured the money we lost not farming there was well compensated for.

Most of the larger swales we drained with field tile. Using

a tile ditcher, we dug a trench about three feet deep through the low wet area and then on out through a low break in the hills. Into the ditch we laid tile, keeping about an eighth of an inch between each foot length section. The tile line was laid on grade, about an inch fall to every fifty feet. Then the trench was again filled with earth. Water seeped down through the ground, into the tile, and flowed on by gravity out to a creek on the other side of the hills.

A line of tile of the four inch in diameter size will generally drain excess water out of the soil about twenty-five feet on either side of it. If the area to be drained is fairly large, as was the case here, the Soil Conservation Service will design a proper drainage system.

Proper drainage is so important, it should not be overlooked even by organic homesteaders with just a small garden. Unlike the swales mentioned above, some land does not look like it "lays wet," although it does. For instance, one of my gardens now lies on a gentle hillside that has fairly good surface drainage. It doesn't *look* wet, and as a matter of fact, can get very dry indeed in late summer. But this ground has a definite drainage problem. The tight clay soil inhibits downward seepage of ground water, and about three feet below the surface, an almost solid layer of rock stops percolation entirely. The effect is wet, cold soil late in the spring. I cannot plant here until about two weeks after well-drained soil in the vicinity has dried and warmed enough for sowing and germination. Moreover, if a hot April day does manage to dry out the upper three inches in this garden, tempting me to plant a row or two, I usually get poor germination because the ground below is still too cold. Also, when a soil "lays wet" like this, heaving from frost is always more severe than in well-drained soil. This garden very definitely needs tile, but unfortunately there is no convenient ditch or creek that I could run a tile outlet to.

If you have drainage problems such as I've described, by all means visit the Soil Conservation Service in your area. Even if you don't qualify for their technical engineering aid, they have plenty of information to help you solve your problem. And you can't enjoy gardening fully until you get your ground properly drained.

As with this one garden plot of mine, one particular swale on the farm posed a problem we could not solve by conventional drainage methods. The hills that surrounded the swale were all too high to cut through on grade with any regular ditching machine we knew of. We would have had to cut a trench that at the highest point would have measured over thirty feet deep. This particular piece of ground encompassed about seven acres, too much, we figured, to "waste" as a wildlife habitat. "Put it in hay," advised an old-timer. "What kind of hay will grow in a swamp?" we asked. "Reed canary grass," he replied without hesitation.

It worked. The reed canary loved the wet land and grew vigorously, forming a matting of roots so strong, we could actually drive a tractor across it without getting stuck, even after a couple of inches of rain had fallen. Reed canary grass is not as palatable as alfalfa hay by any stretch of the imagination, but if cut and harvested before it gets too tall and fibrous, it is nutritious, and cows will eat it. The grass also can be grazed in late summer when other pastures aren't growing.

The steep bluffs falling away to the river valley were kept in woodland, the only possible use for them. Most of the wooded area we did *not* pasture so that by natural reseeding the forest would continue to renew itself. Cattle will eat off new-growing seedling trees.

The sandy light brown soils at the lower levels of the bluffs (where the land again levelled out enough to allow cultivation) faced southeast. Here we planted our earliest vegetable crops—radishes for market, onions, peas, and beets—to take advantage of the micro-climate provided by the sun-facing hills. Here the sandy soil could absorb the full force of the spring sun, while being sheltered from cold northerly winds. Sometimes we had the radishes planted here before the snow melted on top of the bluffs! This early planting not only meant an early harvest, but enabled us to follow up with a second crop (generally soybeans) early enough so that it would mature before frost.

Next, proceeding toward the river, lay the spring-fed swamps. Farmers considered them utter wasteland, but actually they provided us with trout and a winter vegetable, watercress, which grew naturally there. The spring water never froze even in below-zero weather, and our fresh watercress made us the envy of every gourmet in the country. Also, the never failing springs provided extra insurance against dry weather: we could if necessary use the water for a bit of irrigation.

Beyond the swamp was the ridge of black sand measuring about fifteen acres. This land was ideal for berries and for full season vegetables, not to mention the most luscious watermelons and muskmelons I've ever tasted. This field received the best organic care of all—each year we gave it a good heavy coating of manure (as we did also with the light sandy soil). Sandy soils are, by definition, short of organic matter, which the manure supplied while improving the moisture holding capacities of these soils. We rotated crops on this land so a portion was always in clover, which was plowed under for green manure.

The black sand ridge gave way then to a lower level of good silt loam land which reached almost to the river's edge. This land was level, fertile, and deep, and here, with proper rotations, green manuring, and lime, we could farm hard with field crops and not seem to lose fertility. One reason was that almost every other year, the river flooded these bottom lands and deposited another layer of topsoil on them.

By playing nature for all she was worth, we were able to operate a fairly large farm with over a hundred head of dairy cattle, almost entirely without chemicals, and once the farm was in good fertility, our yields were satisfactory, too. We did not have superyields, but then we were not adding to the surplus that farmers were building up at that time either.

I remember one day a salesman stopped by to tell us how much better our corn would grow if it had the nitrogen that an application of anhydrous ammonia would supply—an application he would be more than happy to sell us. We said that he could run his applicator down eight rows of his choice, and if the corn there grew better than the rest, he'd have a customer the next year. He accepted the challenge.

At no time during the growing or harvesting season, could he, or anyone else, see a difference in the corn fertilized with anhydrous. The salesman was astonished. Right on the next farm anhydrous ammonia had a very obvious effect. But that land had been "mined" of nutrients for years.

In raising the fertility of your organic homestead to optimum levels, the first years are the roughest. You must put more into the soil than you take out. That takes a lot of extra time, money, and sweat. Once you have your soil in high fertility, maintaining it organically is not so hard.

My belief is that even organicists rarely get their soil into premium shape. Like myself, they want to, but are limited by time and by a convenient source of organic matter in sufficient quantities. For very small gardens, the matter is not so much of a problem, but when you start working with an acre or more, it takes an awful lot of compost and mulch to do the job properly.

My first move to remedy this situation was to quit making compost piles. Heresy! I hasten to add that compost is the greatest gift you can give your garden. But instead of making compost in piles, I save time by sheet composting—spreading mulch six to eight inches deep right on the garden where it turns slowly to compost and in the meantime, inhibits weed growth and retains moisture.

I could sing the praises of sheet composting for hours. You develop a sort of mystique about the rich, crumbly earth that forms under it. Before I turned to this practice, my garden seemed always caught between mud and drouth. Now, where the leaves lay thick around my permanent bush berry plots I can walk any time of year. The ground is never muddy, never erodes, never moves. The day the thaw is out of the ground and the warm sun shines, I can push back the mantle of organic matter and pick up busy earthworms. But the payoff is in hot, dry August, when less fortunate gardens are

On the Logsdon homestead, composting materials and soil amendments are piled on the garden so any leaching that occurs before they're spread and tilled in bolsters the garden's soil. Saves work in spreading the materials, too.

wilting. My berry bushes look just as lush as if it were June. I can dig down through the leaves, through moist leaf mold, through a layer of moist compost, to a soft, crumbly soil that is actually wet. Across the fence is bare ground, dry as dust.

But I make no bones about it. Concentrated organic gardening is a perfection more dreamed about than accomplished. I have practiced it only on a portion of my acres, a portion which I aim to increase each year. By "practiced" I mean that the ground during every growing season ought to be mantled with either four inches of manure, six inches of grass clippings, or eight inches of hay, straw, or leaves.

That's why I think the first requirement for the organic homesteader is a pickup truck, or if you have to get by cheaper, a trailer you can pull behind your car. That and the ability to enjoy the odd looks you might get from friends and neighbors when you ask them for their leaves and grass clippings.

Better, of course, to get someone else to haul organic matter for you, if you can manage it. If you talk very nicely to your local street department officials (a bit of cash offered won't hurt the situation), you can sometimes persuade them to dump truckloads of leaves on your garden. These are the best leaves, too, because they are usually picked up with a huge vacuum cleaner device which partially shreds the leaves. Shredded, leaves represent much more organic matter relative to unshredded. They also break down into compost faster.

In suburban areas, I have found that institutional groundskeepers must often haul away whole truck loads of grass clippings they mow. If you can intercept them on the way to the dump, they will often donate a load or two to your worthy cause.

But to insure a steady supply of organic matter, you will have to rely on your own truck or trailer. In my area, the general practice is to set plastic bags of grass clippings and leaves out along the road for the garbage collector. In a Saturday afternoon, I can haul as many loads as I have time and energy for. Also, it is easy to get free manure from most people who keep horses, if you do the hauling and loading. If you live in a farming area and can time your arrival at a farm when the farmer is cleaning out his barns with a tractor-operated manure loader, he'll dump a scoop or two right into your truck and save you the work of loading it yourself.

Mulch farming is an art, I think. It has its fine points, not the least of which is timing. Here's how I proceed.

1. I try to get as much mulch-work done in the fall and winter months as possible, before the demands of the planting and growing season are upon me. With leaves, this is easily accomplished, because leaves are only available in late fall.

I put leaf mulch on *permanent* plantings like fruit trees, blueberries, raspberries, blackberries, and ornamentals, especially evergreens. The first reason I do so is because around such plantings you can spread mulch any time. You don't have to wait until the ground warms up and plants are growing as you must with vegetables. Secondly, leaves tend to be acid, but the permanent plantings I mentioned like a little acidity in the soil. Thirdly, the mulch cover keeping the ground around plantings cold in spring is an advantage. The mulch holds back early spring growth sometimes, so that blossoming is held back until after the danger of killing freezes has generally past. This delay in growth would not necessarily be an advantage if I were growing fruits commercially—where the earliest crop commands the best price—but that is not my purpose.

So when I obtain leaves in the late fall, I put them directly and immediately around these permanent plantings. Saves piling and double-handling. I pile the leaves between rows of berry bushes in a sort of windrow about two feet high. I try

There are dozens of materials suitable for mulching. The best are those which decompose and contribute to the soil's fertility, like newspapers and straw.

to keep the leaves away from the immediate base of the plants so the ground there will freeze in winter. (Most fruits do better if the growth cycle is "shocked" by winter freezing, strange as that might seem.) All winter the windrows of leaves make a good place to bury table scraps and garbage. About May 1 (in my area) it is a simple job to spread the windrows out evenly in an eight-inch layer over the entire patch.

You can spread newspapers and rags (I've even used old rugs) on the ground around permanent plantings, then cover them with leaves. Paper is a good mulch, but it's unsightly, and the leaf covering takes care of that.

Another advantage of leaf mulch is that it kills grassy weeds better than other mulches, especially hay or straw. Weed grasses are the bane of mulch gardeners since such weeds can push up through mulch more easily than broad-leaved weeds. Leaves not only inhibit grass growth better than long stranded hay or straw, but also, most grasses dislike the natural acidity of leaves. Where I have a grassy weed problem, I use leaves that have not been shredded. They mat down and smother grass better.

I use straw as mulch for strawberries. But I like strawy horse manure better. I use the latter to mulch *new* strawberry plantings. Then in the fall, I cover the plants with about four inches of plain straw (the strawy manure having mostly turned to compost). The next June I rake the straw off the plants into the middle of the rows. The original manure has decayed completely, and the berries can ripen on clean straw.

Hay is good mulch for certain situations. Spoiled hay is better because decomposition has already begun, and also because you can buy it cheaper from farmers. The trouble with hay is that it often contains weed seeds. In this case, what I do is spread it between vegetable rows and wait until the weed seeds germinate and begin to grow. Then with a fork, I simply turn the hay mulch over, smothering the weeds. I wouldn't call that exactly a recommended procedure, but I'm not too uptight about weeds anyway. A few weeds out in the middle of the row can always be pulled. That's more mulch. Just don't let weeds go to seed in the garden.

Because hay is long-stranded, it's difficult to place it thickly around small vegetable plants without covering them. I confine long stranded mulches to row middles, or as a second layer after the plants have grown tall.

If hay has weed seeds in it, using it in August has another advantage. Many pesky weeds require the long mid-summer days of sunlight to grow properly. In late August they will germinate, grow in a rather stunted fashion, and not go to seed.

Grass clippings are my favorite mulch early in the growing season. They usually contain no weed seed and handle well around small vegetable plants. You can take a sack of fresh clippings and practically pour them along a row of plants. The clippings flow around and underneath small plants rather than bunching up and covering plants like hay or straw will do. Some organicists say that grass clippings mat down so that rainwater won't penetrate them, but I have not found that to be true in my experience.

Clippings rot quickly, supplying nutrients to plants in a hurry. They can also be more easily turned under with a rotary tiller or disk, whereas longer-stranded mulch balls up on the tines.

To spread the workload out more evenly, get your leaf mulching done in the fall. When the ground is frozen, you should be hauling in all the manure, hay, straw that you can get and making handy piles of the stuff as close to where you will need them as possible. I try to put piles all around the edges of my garden plots, so that whatever leaching takes place from winter snows and rains, the nutrients won't be lost, but will run out right onto the garden. (You lose some, of course. Really "artistic" manure farming demands that manure be kept under cover until ready for use, so as to lose the least possible nutrients. For most of us, that would be impossible. You can, however, cover outdoor piles of manure with plastic. I doubt if that's practical though.)

With the piles handy, you can spread them quickly after the soil warms up and plants are growing. If you are handling hay or straw *bales* for the first time, the most practical way to break them up is with a pitchfork. Cut the twine holding the bale together. Sink the fork into the bale, pull a chunk loose, then shake the forkful vigorously. The compressed hay or straw will fall into a fluffy pile. You can also run chunks of bale through a shredder to chop them up, but unless you have a big, heavy duty shredder, that is slow work.

Collecting your neighbors' bagged grass clippings picks up in early summer right when you need them most. Try to get the clippings on the garden right away. If allowed to heat in the bags, the grass packs into a gooey mess that smells to high

Breaking sod or turning in a cover crop is a rough job, and most rotary tillers aren't up to it. This is where that small farm tractor comes in handy. If you haven't one of your own, you can probably hire a farmer to do the job for you.

heaven. It's still good mulch that way, but unpleasant to handle.

On crops like wheat and other small grains that are not grown in rows, true mulch farming cannot of course be practiced. However a coating of manure can be applied to such crops in fall or spring. On small plots of grain that you should raise for your own food (which I will discuss later) you can rotate the grain with vegetables so that the grain has the advantage of the previous year's mulching.

On larger fields, organic matter can be practically applied by green manure crops. All that means is that you grow a grass or legume cover crop to be plowed under instead of harvested. The cover crop can be planted alone, or as a nurse crop with wheat or oats. When the grain is harvested, the nurse crop comes on strong to make a lush stand which can be plowed under the same fall or after it makes new growth the next spring. Rye grass can be planted alone in the fall, then plowed under the next spring. I prefer a legume for green manure, because legumes add nitrogen to the soil as they grow. I use red clover, because in my experience it is better adapted to a wider range of soils.

For plowing under a cover crop, you need a tractor and plow. A rotary tiller is usually inadequate. It almost hurts to plow under a lush stand of clover when you think of the hay you're losing, but the soil benefits immeasurable, and you'll get back the profits of your "lost" hay with increased yields of the crop that follows. I always follow clover with corn.

The organicist's perennial problem of getting sufficient organic matter may be solved adequately if more cities begin to compost their wastes and sewage sludge. I've no experience with sludge, but I have visited with organic farmers who use them advantageously. They use it as a soil conditioner and a mulch around ornamental plants and other crops where the material does not come into direct contact with the edible parts of the plants.

If you have wood ashes, don't waste them. They are a good source of lime and potash. The trick is not to let them lay around where rain falls on them and leaches out the nutrients before you get them on the garden. If you have a fireplace or wood- or coal-burning stove, dump the ashes directly on the garden. I've heard the opinion expressed that coal ashes don't help the soil, but we've made use of them for years.

I generally have huge brush piles to burn every spring—hedge trimmings, orchard prunings, branches of trees cut for firewood. I pile them on the garden and burn them there, then scatter the ashes out. I witnessed a rather dramatic soil response after I first followed this practice on a new piece of ground. I planted corn where I had burned a brush pile. In August, the corn in the immediate vicinity of the burning was more than a foot higher than the rest, with two fat ears on the higher versus one on the shorter. That's how the corn told me the ground needed lime.

Soil fertility is a subject that embraces great complexities, not all of which have been solved to this day. You can spend a lot of interesting time probing into trace elements, or something like iron imbalance or selenium deficiency. If you enjoy soil chemistry, go to it. From the standpoint of practical gardening though, first follow a properly managed, concentrated organic system. Then, if you still have problems—and I'm willing to bet you won't—talk to a soil scientist.

Chapter Five

A Practical Organic Vegetable Garden

The vegetable garden is the heart of your organic homestead, whether you operate a quarter-acre or a quarter-section. There is no reason why you cannot raise all the vegetables your family needs, not only during the growing season, but with proper storage facilities for the entire year.

Vegetables are your most practical enterprise, nutritionally and economically. In the first place, we all know how a vegetable begins to lose flavor and nutrition as soon as it is picked. Pure logic tells you that no matter how streamlined the commercial operation may be, a time lag of a couple of days at the least elapses between harvest and the time you get the items home from the supermarket and onto the table.

As for economy, the price of fresh fruit and vegetables is high now, and there's every indication it will go higher in the future. Each price increase means that the time you spend in your garden is rewarded that much more.

Even if machines take over all the harvesting of fruits and vegetables, your home garden will still reward you in special quality ways. Varieties of fruits and vegetables developed for machine harvest often lack quality of texture and taste you find in homegrown kinds. For instance, the big, lush, juicy tomatoes you can grow can't be handled by mechanical tomato harvesters—the machines would mash and bruise them.

As you approach the project of seriously raising all your own vegetables, be practical. That is, stick to vegetables that have proven dependability in your area, and don't waste time on the rest. As an organicist seeking to avoid the use of harmful chemicals, you start out with the odds against you. Don't let anyone kid you about that. Successful organic gardening is not easy, and it has no magic formula to shoo away bugs and blights. Rule number one: Plant varieties with known resistance to disease, and stick to the dependable kinds of vegetables. Remember that the veteran gardener whose skill we praise has often simply learned over the years which plants to avoid growing in his garden. He has followed the organicist's byword: work with nature, not against her.

Rule number two for successful vegetables: Do *not* plant too early. Much better to try to extend your gardening a month later into the fall than two weeks earlier in the spring. Early planting means risky germination. Early planting, if germination does occur, means rotting seed if coldish weather lingers, or slow stunted initial growth of seedlings. Early planting means bugs, slugs, and cutworms when vulnerable, tender young plants have no vigor to counter insect attacks with fast growth. Early planting feeds birds and rabbits when they are hungriest. Early planting means a weedy garden.

Another general observation. Rabbits, squirrels, groundhogs, crows, blackbirds, and raccoons, if numerous in your neighborhood, are going to make gardening dreadfully frustrating at times. We live in a fairly heavily populated area (people), but there are more wild animals around than when I lived in very rural areas of Minnesota.

Rabbits will go after almost any tender new growth after

a winter of chewing bark and shoots. They especially like new growth on young raspberry plants and peas. If they can't get peas, they'll eat lettuce and carrot tops. And sometimes stringbeans. I fence my peas in as a matter of course. Carrot tops seem to come back of their own accord after being eaten, and rabbits don't seem too fond of Buttercrunch lettuce.

I shoot rabbits. Criticize me all you want. If I didn't, there wouldn't be a live plant in the whole neighborhood.

I'll mention other specific pests and specific remedies when I discuss particular plants in detail.

There are three kinds of vegetables from the standpoint of the mechanics of planting. 1.) Those you plant "once in a lifetime"; 2.) direct-seeded vegetables; 3.) those you start indoors and then set out in the garden. I shall discuss these three kinds *in the order you should plant them*—or as nearly as I can come to that schedule.

Rhubarb

I like its old name better—pieplant. I've always gotten my rhubarb "starts" from a neighbor. Very early in spring, I dig up an old plant, separate the roots, and re-plant each separately.

Pieplant is a perfect vegetable for organic gardeners. I've never seen anything bother it, possibly because the leaves are poisonous. Rhubarb just keeps on growing forever if you put a little manure around it every other year or so. Rhubarb stalks are the first new-growing vegetable I harvest in spring. Its value lies mostly in that fact—there's nothing else growing at that time.

Rhubarb freezes well. Don't cut stalks after hot weather arrives. Cut off blossoms if they appear, so the vigor of the plant goes into the root, not into seed.

While freezing is more practical, rhubarb can be forced the old-fashioned way for winter use, if you have extra plants.

You dig up roots in the fall, leaving as much dirt on them as you can. Let stand on the ground until after a good freeze. Then put the roots into a box of sand and set them in a dark cellar where the temperature is around sixty degrees. Keep the sand moist. Very tasty stalks will grow up in a couple of weeks. Afterwards, the roots are used up and won't grow again. Throw them away.

Asparagus

This is the other vegetable you need to plant only "once in a lifetime." It merits high rating for the organic homestead. It's one of your "musts." First of all, it's only good when rushed five minutes from garden to steaming water to table. Secondly, it responds superbly to organic fertilization. Thirdly, once planted and cared for lovingly, it will last as long as you will. Fourthly, it requires little labor to grow. Fifthly, it is ready to eat early in the spring before your seeded vegetables are mature enough for the table. And sixthly, asparagus is a good item to sell "by the little" because freshness is so important and so hard to find in a store.

Asparagus won't produce a crop for three years if you plant it from seed, for two if you start with year-old roots. But once it starts producing, it will send up tasty spears year after year. Plant it in a quiet, sunny corner of your garden with lots of room for root growth, manure it well, and plant a cover crop of cowpeas or soybeans between the rows after the spears are harvested.

The "proper" way to grow asparagus is to buy second year roots (one year old roots). Place them in a hole or trench about six inches below the surface of the soil, being careful to spread the roots out, fan-shaped, in the bottom of the trench or hole. Do the job as soon as the ground is workable in spring. Cover with no more than two inches of soil. As the stalks start to grow (they will be quite tiny at first), gradually fill in dirt around them until the ground is level. Do not harvest any stalks the first year, nor the second. By the third year, you should be able to cut stalks until hot weather—or for about a month from when the first stalks appeared.

I have to be a little obstinate and raise asparagus my own way. I start my plants from seed in peat pots full of vermiculite. After the seed grows to a plant of four or five inches, I set the plant, peat pot and all, in a six-inch hole, then treat it just like plants grown from roots. By the third year after planting, I get almost as many spears to eat from my plants as from roots planted at the same time.

If you can dig up your own roots from an old garden, with clumps of dirt attached, you can take a cutting a whole year sooner. The roots you buy have been out of the ground awhile, washed, and most of the tiny root hairs are gone. The shock they have gone through sets them back so that they grow to maturity only a little faster than plants grown from seed. Lots of people might argue that, but I've tried it both ways. If you do purchase bare roots, by all means discard the spindly ones.

I am not going to tell you to plant asparagus (or anything else) in "fertile, well-drained soil." That's what it says on the seed packet—and in more books than I care to believe. It is unctious, witless advice. Every vegetable I grow prefers fertile, well-drained soil. How about yours?

I use chicken manure to fertilize my asparagus. Asparagus is a glutton for nitrogen, and chicken droppings are a rich, natural source of it. I spread a very thin layer of the manure early in March before growth starts (thin enough so the layer does not insulate the ground from sunlight and keep soil cold). Towards the end of the harvest season, around June 1 for me, I put on a six-inch layer of chicken litter, effectively killing any weeds that have begun to grow. If weeds are extra bad, I cultivate or hoe before adding the mulch-manure.

In fall, I grind up the stalks with my Gravely lawn mower (the circular mowing blade on this machine does a good job of shredding, too) reducing the plants to mulch. Every third or fourth fall, I add a coat of lime to the asparagus patch. My soil is naturally a little acid, and asparagus likes a neutral soil.

Asparagus has two bug enemies, the asparagus beetle and the spotted asparagus beetle, upon which I annually exhaust my supply of cuss words. You can hand-pick the pretty little devils if you have only a few plants. Rotenone helps some. (I buy a big can of rotenone powder every year, and use it *liberally* whenever a bug problem comes up. Rotenone is not a very effective insecticide, but it's one of the few we crazy organicists can use with a clear conscience.)

The spotted asparagus beetle first appears as an orange bug about the size of a firefly. The plain asparagus beetle is the same size but has more black and less orange in its coloring. If you leave either one to its own devices, pretty soon you will find olive-colored larvae about a quarter-inch long merrily chomping away on the plants.

Chipping sparrows eat them, but not nearly enough. Chickens, too, if the dumb clucks happen to see one before they see a ripe strawberry. Fortunately, the beetle doesn't get bad, at least in my patch, until after the cutting season. At that point, you have only to control them so they won't weaken the fern-like plants too much while the latter are storing up the food in their roots that makes next year's harvestable stalks. The best defense at this time against the bug is a well-fertil-

Before planting or transplanting in your garden, it's a good idea to work in some organic fertilizers, like rock phosphate, lime, greensand, and blood meal.

ized plant, which will grow vigorously enough to handle the beetles you didn't kill or chase away with rotenone.

We figure ten plants for each member of the family will supply enough stalks for eating fresh during May and for freezing for winter use.

Direct Seeding vs. Transplanting

After rhubarb and asparagus, the rest of the important vegetables have to be planted every year. The first question is: Which should I plant directly into the garden, and which should I start indoors? My answer: Start indoors only tomatoes, sweet potatoes, peppers, and eggplants. Many skilled gardeners start head lettuce, cabbage, cauliflower, broccoli, Brussel's sprouts, melons, and so on indoors. If you live far to the north and are equipped for the task, go to it. My experience is that much transplanting is totally unnecessary and that the modern homesteader has enough to do without unnecessary work. Except for the four crops I mentioned, wait till the ground is right, then seed direct. Proceed in this order.

Onions

The first things I plant in the spring are onion sets. They're the only things I'll plant before the ground is fit. You can plant onion sets right in the mud, if you want a few very early. I generally take the hoe and dig a trench about four inches deep, stick the onions firmly into the bottom of the trench—root end down, of course—then close the trench. If the ground is muddy, don't pack it. Unless exceptionally cold weather follows, the onions soon make scallions.

I plant some every week or so until I have four different plantings. This keeps us in fresh scallions until July. The ones that get too big make good creamed onions. This is not the proper place in your planting schedule to discuss it, but I also plant a row of onion *seeds*—same time as I plant cabbage. Onions from seed keep better for winter use. And with seed, you have a much wider selection of varieties. Plant seed very shallow—just barely cover. Use thinnings for scallions.

The organic gardener should get to know onions well. Other than the onion maggot, which is no problem in my garden, the onion has no enemies. It is foolproof and easy to grow. And good—raw, cooked, or French fried.

When pulling or digging mature onions in the fall, let them lie in the sun a day or two to cure before storing them.

Potatoes and Peas

Some traditions say to plant peas and potatoes on St. Patrick's Day; others claim that Good Friday is the proper time. My advice is: Observe religious days in church, not in the garden. I've seen too many Good Fridays when snow covered the ground. Spuds on St. Pat's Day may be just fine on the

Emerald Isle, but in my part of the country, they are liable to have their noses frozen if planted that early.

Peas and Irish potatoes do like cool weather and will germinate and grow when the soil is still too cold for some other vegetables. But don't carry that notion too far. Especially with peas. The earliest pea varieties (the smooth-skinned peas) will stand more cold weather than the wrinkled pea varieties, but the taste of the former isn't as good. The wrinkled peas—the ones you should plant—can go into the ground as soon as the soil is dry enough to work—*don't mud them in.*

If you want to eat well from your garden, not just win some kind of race with your neighbors, don't be in a hurry to start planting. In my area of southeastern Pennsylvania, the ground almost invariably dries up enough to cultivate right around the first of April. The temptation to plant is almost irresistable. Those with sandy soil can give into temptation and perhaps fair well, but for me, with clay soil, it's a good time to take a trip away from home. If I yield to temptation, which I will do if I hang around the house, and plant some peas, we will get our usual run of April weather—snow, cold drizzle, temperatures in the low thirties, followed by one last thumping good freeze. You might as well wait. Even if unharmed, the seeds aren't going to flourish until warmer weather anyway.

Planting potatoes requires little skill. I like to say the reason the commercial potato industry finds itself so often in the doldrums of low prices and surplus crops is because it is so easy to grow spuds.

Trees may need God to bloom and bud,
But any fool can grow a spud,

as one melancholy Maine potato grower once sorrowfully intoned.

Don't buy "seed" potatoes unless you are buying good certified seed. Often, at your local farm and feed store, or other

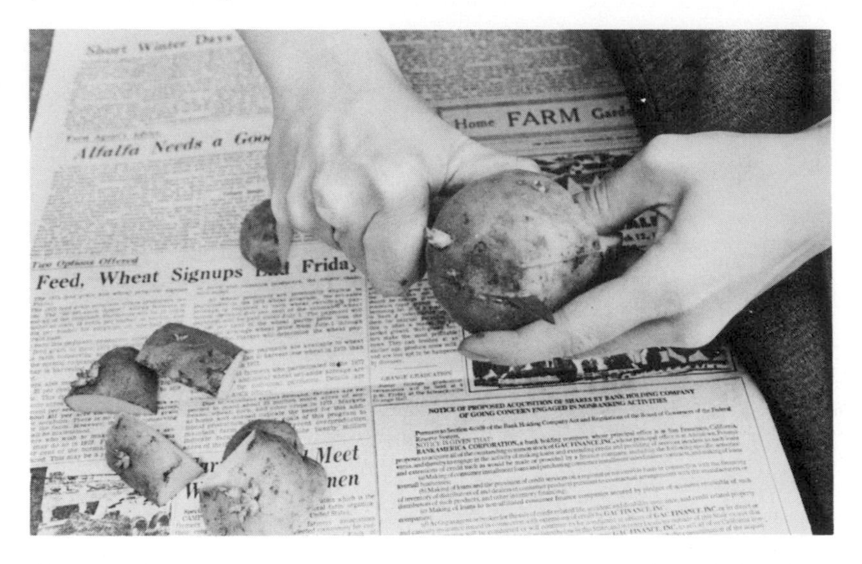

Potatoes that have started to sprout can be cut up and planted. Just keep a piece of potato with each sprout. Cheaper than seed potatoes.

general purpose garden store, "seed potatoes" are runts culled out of good big spuds. Better to buy ordinary eating potatoes from your grocer, if you need only a few. Be sure the potatoes haven't been treated with a sprout inhibitor. About the only way to be sure about this is to wait for them to sprout. What I generally do is use potatoes that my wife has bought for the table but not used and now wants to throw out because they're sprouting away in the bottom of the sack.

My best results follow planting of medium *whole* potatoes. But good plants grow from slices of medium or large spuds. Each slice must contain an eye or two.

I plant the potatoes about four or five inches deep, about a foot apart in the row and about thirty-eight inches between rows. I cultivate once, then lay on the mulch. When potatoes develop, they sometimes push up out of the dirt and get scalded by the sun. The mulch prevents that.

Potatoes respond very well to mulch anyway. They like cool, moist weather, and the mulch keeps the ground shaded and damp around them.

The Colorado potato beetle likes your plants even better than you do. You can control him by handpicking on a small patch—I do on two rows 100-feet-long each. The secret is to get the first few bugs that arrive. Never let the little dears proliferate. Dust plants with rotenone if it makes you feel any better.

I don't know what varieties to tell you to use. Whatever is handy. Experiment. Find out which ones do best for you.

Be sure to "steal" a few new potatoes out of the hills when you have your first peas. If you dig gently into the side of the potato hill, you should be able to get a handful of small ones

Peas, especially sugar peas which are eaten pod and all, should be picked before they reach full size. They taste better picked at a young, tender age.

with out harming the plant. With the peas, they are very tasty. In fact, fresh potatoes out of the garden any time are good. You may have forgotten, or never known, just how much tastier a truly fresh spud is, compared to those that have been stored a long time.

Potatoes need a dry, cool, dark storage bin, kept at about forty degrees with good air circulation. However, I "store" mine right in the ground where they are growing. They'll last to frost time. With a covering of mulch, they'll last on until Thanksgiving in a normally dry fall.

During winter, you can store potatoes in boxes of dry sand, with hay bales ranked two deep around the boxes. You can store any root crop that way.

Peas do not have many bug enemies, but rabbits love the new vines. I pound steel stakes in the ground around the pea patch, unroll a lenth of chicken wire, and fasten it to the stakes. Doesn't take long, and saves a lot of grief.

Pick peas before they get tough. They taste much better if harvested at three-fourths full size, even though that means it takes more for a meal. Shelling peas is slow work, which is the reason we do not freeze as many pints as we do of some other vegetables. But all the work you do on peas is worth it many times over, when you taste that first mouthful ten or twenty minutes from the garden.

During the next warm spell after you get peas planted or a week to ten days later, plant the crops listed below. If you notice a few missing, like cauliflower, I'm doing it on purpose. I am telling you how to plant dependable vegetables. If you want to try some of the others, more power to you. The vegetables I don't mention I've quit growing seriously for one reason or another.

All six vegetables are planted about the same way. A fine seed bed and very shallow planting—a half inch is plenty.

Lettuce

An especially reliable vegetable for the organic gardener, because bugs don't hurt it much in the home garden. Sometimes outer leaves get eaten a little, but the tenderest inner leaves of butterhead varieties usually remain uscathed.

I used to grow every type of lettuce, but now I plant only bibb, and more specifically, the butterhead bibb variety called Buttercrunch. To me it is far superior in every way, so why grow anything else?

A row of lettuce always goes between two rows of broccoli or between broccoli and cabbage—anything that grows big and spreads out a lot. By the time the broccoli closes over the row, the lettuce has all been eaten or grown too old for table use.

Radishes and carrots

I mix the two kinds of seed and plant them together. The radish seed mixed in prevents me from planting carrot seed too thickly. The radishes come up very fast and mark the row, so you can cultivate sooner without plowing out the slower-germinating carrot seed.

As you harvest the radishes for the table, the space they occupied becomes available for the spreading carrots. You can start eating the radishes when they are the size of marbles. Start on the carrots as soon as they reach the size of your little finger. Not only do they taste best at that size, but if the stand is too thick, you give the other plants more room to grow. We just keep using carrots out of the ground until cold weather, at which time I cover some with a heavy coat of mulch. We've eaten carrots stored this way on New Year's Day, dug from the ground under a foot of snow.

Since spring-planted carrots get rather strong by late fall, it is best to start another planting around July 10 if the ground is not too dry. We over-wintered some July-planted carrots under mulch, and dug them up in March, still in fine condition.

Other than rabbits occasionally nibbling on carrot leaves, I experience no insect or disease problems raising carrots or radishes organically.

Broccoli

Two fifty-foot rows of broccoli will supply a family of four more than enough of this vegetable fresh and frozen. Sprinkle the little black seeds along a very shallow furrow. Just barely cover. When the seedlings are about an inch tall, go through the row with a hand weeder and take out all weeds. Thin broccoli so they are at least a foot apart. I generally douse the little plants with rotenone at that time.

Nothing much else to do to broccoli except eat it. Keep the bud heads picked if you want the plant to go on producing more.

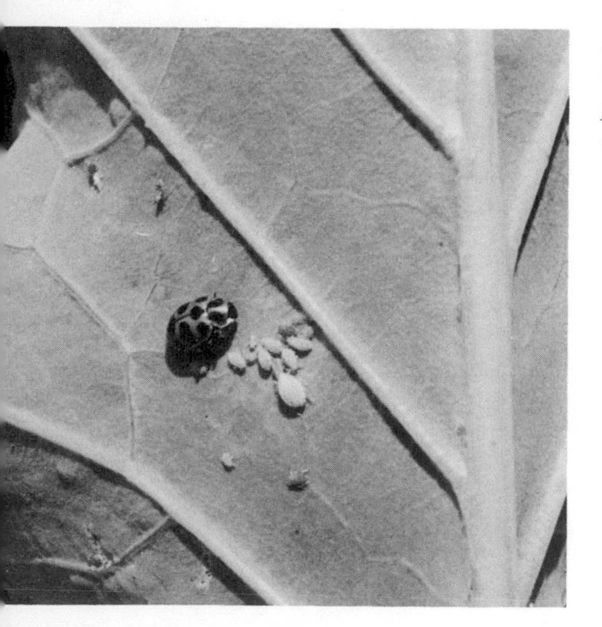

Ladybugs dine on aphids and many other plant pests. You can purchase ladybugs by the pint to help protect your vegetables.

Aphids bother broccoli. Actually, the little plant lice bother *me* more than the broccoli. They seem to do little harm to vigorously growing plants, but they are difficult to clean out of the heads you bring into the house to eat. Cookbooks prescribe soaking the heads in vinegar, which is supposed to convince the aphids to move out. Didn't work for us.

I do not know how to control aphids on broccoli. Fortunately, the pest does not seem to affect every plant. There may be a colony on the first, then two without any, then another bunch on a fourth plant. A mystery. Ladybugs eat aphids, but not nearly enough, in my opinion.

Beets

I always do well with this vegetable. Beets are really two vegetables in one. The tops make excellent greens. If you get beets planted too thick (and that's not hard to do because each of those little hulls in the seed packet can contain more than one seed) thin out the stand as the beets reach marble-size and fix the whole plant as greens. Waste not, want not.

Golden beets taste about the same as red ones, but don't stain as badly. Both kinds seem to grow equally well.

Cabbage

If you can grow only one vegetable of the cole family, make it cabbage. I highly recommend that you try Savoy. Maybe that is a matter of taste, but I must have lots of company. I notice that when I do see Savoy in the supermarket, it is priced about twice as high as other varieties. Worth it, too.

Savoy is easy to recognize by its characteristically wrinkled, crinkly leaves. Most seed companies carry at least one type of Savoy. I favor Burpee's Savoy King.

The cabbage makes big heads in good soil. In fact, all too often it makes heads so big they burst at maturity, especially after a heavy rain. When our cabbage heads split, we pick them and make sauerkraut.

Cabbage worms usually make their appearance about the time the cabbage heads begin to form up if not before. For some reason, I don't have many cabbage worms. I throw a lot of rotenone around when I see white cabbage butterflies dancing down the row. Wood ashes have been used to thwart the cabbage worm for almost a century. A newer organic control called Thuricide is supposed to be especially tough on cabbage loopers, though I haven't tried it myself.

With cabbage in the ground (plant it just like you plant broccoli and other cool weather early vegetables), wait another week before starting hot weather crops. The first of these is corn. But I will talk about it in the chapter on grains.

Beans

About a week after you plant your first corn, it's time to sow snap beans. You can't call them string beans without angering the commercial growers who insist string beans don't have strings anymore. It's a crazy world. The bean I love best is Kentucky Wonder Pole Bean; it does have strings, and I don't care.

It is futile to argue taste, I know, but I am going through life worried that many people will never experience the best tastes. So I will just have to tell you that no bush bean or snap bean in the world can stand up to the taste of a pole bean, French gourmets and cooks to the contrary.

For us, the Kentucky Wonder is an all purpose bean. You can eat them very young and tender, like a true gourmet (don't buy all that young and tender business till you've eaten a full-grown pole bean from a pot containing a slab of smoked ham); you can eat Kentucky Wonders medium-sized or large after you peel the strings off; and you can shell out the fully matured beans and use them as baked beans. All three ways the beans are delightful.

Since we are very much a bean family, we do plant a few

bush beans right after first corn planting and before the pole beans go in. Bush beans can stand some cool weather better, and they produce beans a whole lot quicker than the pole types.

Plant pole lima beans a week or ten days after pole string beans. Not any sooner. Plant a second crop of pole string beans ten to fourteen days after the limas. (Plant some soybeans, too—a subject I'll discuss in the chapter on grains.) That should give you beans enough for half an army.

Beans, snap or lima, bush or pole, have about the same enemies: slugs when the plants first come up—especially when you plant too early. Cutworms, too. A paper guard around each stalk is supposed to stop cutworms, but that is rather impractical for anyone growing more than a dozen plants, which you certainly will be doing. Cultivation discourages slugs and cutworms. When you see where a cutworm has been working, you can often dig in the surrounding soil and unearth the rascal. A shallow saucer of beer, say experts, will lure slugs to it and drown them. Didn't work for me.

The bean's worst enemy in my garden is the Mexican bean beetle. He starts to work about the time the beans begin blossoming. He eats a few holes in the leaves, lays a batch of orange eggs, and pretty soon a whole colony of orange-yellow larvae appear. I don't know any really effective way to stop them organically, other than handpicking, which can be difficult. I've planted marigolds between the bean rows, and that helps, but only a little. I douse the vines wth rotenone, and that helps, too—a little.

Fortunately, we've always gotten enough beans before the beetles ate the plants up. But we'd surely get a lot more, especially of dry beans for baking at the end of the season, if the Mexican bean bettle would go back to Mexico.

You plant all beans about the same. Seeds about three to

four inches apart in the row. Plant seed about an inch deep, if the weather has been rainy; up to three inches deep in dry weather. Plant limas with the eye down; they'll come up faster.

Pole beans demand the extra work of making and setting up the poles. I cut my poles during the winter time, a by-product of cleaning brush and volunteer trees out of my

Using poles culled from your homestead's woodlot, you can easily construct a pole bean trellis like this one in the Logsdon garden. A wire stretched between two rows of beans supports the trellis pieces.

hedgerow. At planting time, I work the bean ground up with the tiller *first,* then put up the poles.

My pole construction amounts to a two-row trellis. First I set posts in the ground at either end of the twin rows to be planted (rows about 38″ apart). I stretch a wire between the posts. Then I set the two rows of poles in the ground on the outsides of the two rows of beans. I shove them into the ground with all the strength I can muster, then lean them in on the wire (from both sides) and tie them tightly. The effect is a long tepee-like frame straddling the two bean rows. With the hoe, I then mark out my two rows under the poles and plant. The beans come up and the vines wrap round the poles. Seeds do not have to be right next to the pole.

Poles should be about ten feet long. Mine are too long, and I have trouble reaching beans that develop at the top of the poles.

Keeping the beans picked will encourage more to develop, unless the bean beetle is bad. When you have used all the beans you need for fresh and frozen, let the rest mature on the stalk for your dry beans. In the fall, shell them and store in the refrigerator, if possible. As with weevil in wheat, there's a bug that can get into dried beans where it multiplies and eats holes in the beans, ruining them. Refrigeration keeps any eggs that might be present from hatching. You can also heat the beans to kill the eggs—as I explain in the chapter on grains.

Limas are the least dependable of the beans under discussion. They don't like cool weather early, and they don't like real hot weather later on. Extremes of temperature or humidity—either way—cause blossoms to drop off rather than form bean pods.

Beans are a legume and don't like acid soils. Lime before beans. You can also use a legume inoculant available from most seed stores. It's a black powder that increases nitrogen activity and, therefore, can boost growth and yield.

Cucumbers and Squash

After your first planting of beans, put in cucumbers and squash. For cucumbers and *winter* squash, plant in "hills," about five to seven seeds per hill. Hill does *not* mean a raised mound. Pull out all but three of the most healthy plants after the first true leaves begin to grow. The hills should be spaced at least four feet apart in all directions. Five or six wouldn't hurt.

I plant summer squash in rows and thin to one plant every three feet.

Cucumbers can be grown on a trellis if you are hard up for space. Sometimes I plant cukes next to early corn and let the vines trail into the corn stalks to save room. But only in early short corn that does not shade the ground too much. I don't believe in planting vine crops of any kind right in the corn rows, even though I've read that other people claim good luck doing that. My experience is that planting vine crops in corn cuts yields, because the corn shades out sunlight and gobbles up all the soil nutrients.

Cucumbers suffer from wilt in my garden, so I must buy wilt resistant seed. M & M from Burpee has proved the best so far. I recently read in an old gardening book that wilt can be prevented by keeping the stalks at the base of the plants covered with soil as the plant grows. I will have to try that. Another trick from an old book says that wilt can be controlled by burying a small length of cucumber or squash vine out about two or three feet from the base of the plant. The buried section takes root and keeps the ends of the vine growing more vigorously. I intend to try that too this year. (Incidentally, an organic gardener should make an effort to collect old gardening books. After all, it may be hard for chemical companies to believe, but man was raising fruits and vegetables long before the advent of modern sprays. When you visit

Zucchini is a favorite summer squash. The best way to plant it is in hills. Scoop a hole into the garden soil, drop in some compost and a half-dozen seeds, and close it up. That's a hill. Each should be located at least three feet from any neighbors.

book stores and antique shops, you can sometimes spot old books that contain good, organic information.)

The worst enemy of cucumbers and squashes is the cucumber beetle, a black and yellow striped bug about half the size of a firefly. The beetle is especially harmful when the plants are first coming up and when the weather is cool, holding seedlings back from fast growth (another reason for not planting early). A strong tea made from tobacco and drenched on the young plants can help. So can a mixture of ashes, lime and rotenone. If you have cigarette or pipe smokers in your house, save the contents of ash trays for fighting beetles. Pipe ashes, with unburnt tobacco mixed in, are the best; you have to break up the cigarette butts. If it rains, get out there right afterwards and repeat your ministrations. The beetles can ruin a seedling overnight.

The only foolproof defense against the cucumber beetle is to cover the hills with protective caps of window screening before the seedlings come up. Make sure the edges of your caps stick down into the ground. And sprinkle rotenone generously around those edges. You can make protective caps out of old window screens, or you can use old-fashioned hot caps. Once the plants are growing well and warm weather arrives to stay, the worst danger from cucumber beetles is passed, in my experience.

Squash also suffer depredations from the grey squash bug in some areas. The squash bug does not attack until the plants are growing well, and therefore is not a critical menace. It prefers winter squash to summer squash, and seems to prefer acorn squash to butternut. For this reason, and others, I recommend for winter squash, only butternut varieties for serious organic gardeners.

Old-timers used to lay boards among the squash vines, under which the squash bugs would congregate at night. Early in the morning one well-placed stomp under a lifted board got rid of a lot of bugs. A tobacco tea or powdered tobacco leaves will deter squash bugs a little.

For summer squash, I grow yellow crookneck and zucchini. Dry weather will reduce yields in a hurry. Both kinds, but especially crookneck, respond to the extra moisture-saving properties of a thick mulch.

Tomatoes, Peppers, Eggplant

Now is the time to be setting out these three vegetables that you started indoors a month ago. Set out tomatoes first, after you are sure all danger of frost is past. Even a light frost can harm tomatoes, but the plant is tough in other ways. I've seen tomato plants sit in muddy ground and almost turn blue in forty degree weather, then snap out of it quickly with the advent of hot weather. One reason is that the tomato has few bug enemies during those first critical days in the garden.

Five days after you transplant tomatoes, you can put eggplant and peppers in the garden. Peppers are more tender to cold weather than eggplant, but both need a long hot season to produce well.

If you have a sandy soil or a light loam, you can plant all three vegetables by direct seeding in the garden and get a crop. But your chances are better, if you start them indoors. You can monkey around with hot beds and cold frames if you perfer, but I think that's a waste of time. I've had good luck with an easier method.

For a family of four, you'll want about twenty tomato plants, eight eggplants, and eight pepper plants. You'll need four flats measuring about sixteen inches square to start that many plants. Into the flats, put about an inch of vermiculite, available at all garden stores. Buy as many Jiffy 7 expandable

A quick way to get plants started in flats involves the use of expandable peat pots surrounded by vermiculite. The flats can be put in a sunny window or a cold frame as soon as the seed sprouts.

peat pots as you need—they're little round wafers when you buy them. Place the wafers in the flats on the vermiculite and sprinkle water on them. They swell up into little pots, about two and a half inches tall. Then fill the space between the pots with more vermiculite. Poke two seeds into each pot, no more than half an inch deep. Then put the flats in a warm place until the seeds germinate. Keep moist, but do not oversoak the pots. I put my flats on top of the furnace. As soon as the little seedlings break above the pots, put the flats in a sunny window. Keep watered, but don't drown the poor things.

The main problem with growing these plants indoors is that it is sometimes difficult to get enough sun for them just from a window. Supplemental light from flourescent bulbs solves the problem. Garden stores sell special stands equipped with bulbs to grow plants under. They may seem expensive, but surely a lot cheaper than a greenhouse.

Without this equipment, I put my plants outdoors every day the weather is warm enough. But I cover the plants with old window screen to shade them partially. Straight sunlight, before plants have "hardened off," can kill them, or turn them a pale whitish color. Gradually decrease the amount of time the plants are under the screen until they are used to direct sunlight. Then they are ready to transplant to the garden.

All three vegetables are transplanted the same way. Carry the flats to the garden. Set the plants, peat pot and all, into four-inch-deep holes you have dug in the soil. You can set tomatoes deeper than they are growing in the pots, and should, if the plants are leggy. You want the tomato plant anchored firmly in the soil so it won't sway in the wind.

Give tomatoes a lot more room than eggplant or peppers, unless you stake them. Staked tomatoes with suckers kept pruned are supposed to produce nicer, larger fruit, but I think it is a waste of time. Mine run all over the ground without stakes or pruning, and I get gobs of nice, large tomatoes.

After the weather is good and hot, I mulch the tomatoes and eggplant with manure if I have it, with hay if I don't. Usually, I mulch manure close to the plants, and hay out farther, so the ripening fruit lays on the hay, not in the dirt.

The tomato should be the organicist's favorite vegetable. If you use wilt resistant varieties, about the only other pests are a few big green tomato worms, easy to control with hand picking. Hold a jar in one hand, and brush the worm in with a stick in the other. The worms chew off leaf clusters and you can usually spot their handiwork before you spot them. When you see just stubs of branches leading off from the main stalk, know that a worm is near.

Tomatoes are very versatile. What you don't use fresh,

Tomatoes are practically everyone's favorite. They benefit from a mulch —sawdust used here—and can be staked or tied to a trellis.

canned, or frozen, you can make juice, chili sauce, and catsup out of. When frost is expected in the fall, you can pick green tomatoes for pickling or frying. If they are beginning to turn red, you can pick and store them in the house where they will ripen. Some people wrap these late tomatoes individually in newspaper to ripen, but it seems to me they do just as well on a shelf, unwrapped.

Peppers are another winner for the organic garden, because at least in my experience, few pests bother them. Remember, you can pick peppers green, or after they turn red.

Eggplant gets speckled with very small holes, done in by a small black fly. I douse the leaves with rotenone. With a good manure mulch, the plants grow so fast after they get started that they produce a crop despite the bugs. But eggplant is not really one of the organicist's dependable vegetables. I think I grow them mostly for the unusual and beautiful purple color they inhance the garden with.

Sweet Potatoes

After your peppers are in the garden, you can safely set out sweet potato plants. I start mine in the house about a month before setting out time. I select three sweet potatoes from those my wife buys at the grocery, and plant them in a bucket in six inches of sand with three inches of manure under the sand. I keep the sand moist, and soon the three potatoes sprout many plants.

You can gently break off the shoots from the potato and plant them, making sure you water them the first couple of days after setting out. The plants will wilt at first, but generally they come back. Though it is not supposed to be necessary, I allow a part of the potato to go with the sprout into the ground. I can usually slice the potato into four pieces, each with a healthy plant on it. These seem to start growing a little quicker.

There are several kinds of sweet potatoes, but only one kind is worth growing, for my taste: the deep yellow, *moist* varieties that some people persist in calling, erroneously, Louisiana yams. The growing season is not long enough in my area to grow yellow, moist sweet potatoes commercially, but I get enough.

I encourage you to try sweet potatoes. They don't seem to have bug or blight enemies at all.

Pumpkins

I plant a few pumpkins late—around July 1—so that they are ripening closer to Halloween time. Earlier pumpkins sometimes rot before we can make jack-o-lanterns out of them. If you want to grow those great big pumpkins, plant them earlier, and handle them the same as you handle squash.

Endive

By now your garden should be at its zenith, and the spring and early summer plantings growing to beat the band. In my area, the time is July 4 to 10, the time to plant that fall garden. I tear up the pea patch, and the first thing that goes in there is endive. Plant it just like you do lettuce and with luck you will have a green salad until December. I say with luck, because crucial to a successful fall garden is a good rain in early July to get the plants off and running.

You are supposed to blanch endive before picking it, but I don't. Blanching is whitening plants—to eliminate bitterness—by covering them or tying their leaves up to shield them from light. Endive is one of several vegetables that are commonly blanched just before picking.

Though I don't purposefully tie up my endive to blanch it, I do plant it very thick, and the crowded plants blanch their inner leaves a little anyway. However, I happen to like endive blanched or green. I sometimes cover my endive in late fall

By covering endive with these plastic panels—bent a bit and tied with string, the panel holds the u-shape—the Logsdons are able to enjoy it garden-fresh even when there's snow on the ground.

with some old plastic panels I have. We've enjoyed endive from under the panels in winter when there was a foot of snow on the ground.

Kale

I call kale the wonder vegetable. I wonder how it keeps on growing right up until very cold weather. Kale is rich in vitamin C, not readily available from the garden in November and December. Moreover, if your winter is not too harsh and you give kale a little protection, the plants will start to grow again in early spring and provide you with greens even before asparagus is ready.

Scotch curled kale is the best tasting, but it won't come up as dependably as Siberian, especially when planted in July. So

I have gone to Siberian. The first big leaves of Siberian should not be picked. Wait until the more curly leaves come along.

To enjoy kale you need a wife who was born in Kentucky and raised on the stuff. She knows how to cook it with a slab of smoked ham and onions into a meal fit for kings.

Aphids bother kale, but not much. A good hard frost will usually (but not always) get rid of aphids but it only makes the kale taste better.

Turnips, rutabagas, kohlrabi

All three make dependable fall crops. I mention them for that reason. The reason I can't get excited about them is that about one meal of each is all my family cares to eat a year. You can always feed them to the cow, if you have one. Turnips are such an exuberant vegetable; even planted late, they make huge crops. So I plant them to give the garden the look of life in drab November. The leaves of turnips make delicious greens, too, but we are always stuffing ourselves with kale about that time.

During the first week of July, I also work up the old strawberry patch, mowing and shredding the plants with the Gravely, then rototilling the ground. Here I plant a late crop of corn, another row of carrots, and perhaps some parsnips.

Parsnips are not, perhaps, the best tasting vegetable in the world, but they are easy to grow and winter weather doesn't hurt them. In fact, freezing makes them taste better. Early in spring, before any other crop is available—even in February and March—I can dig up a meal of parsnips. This old root vegetable rounds out my schedule—to have something to eat every month of the year from the garden.

A final suggestion about vegetables. Only experience can tell you how much of each kind to plant so you will have an adequate supply for eating fresh and for freezing until the next season comes. It will depend on your family's tastes more than anything. But whatever you do determine as the right amount to plant—*plant a little more than that.* As an organicist, you may have to sacrifice (or share) some of your produce with the bugs and the blights and the animals. Planting more than you need means you may get enough in a bad year. Of course, in a good year you may have enough to feed the whole neighborhood.

A final, final suggestion. I've said little about tillage, and the reason is that tillage practices vary with soil type. What I do may or may not apply to you. In general, experience has taught me that the organicist who uses a lot of mulch will benefit most by turning his ground over in the fall with a plow. A rotary tiller would incorporate the mulch and litter and green manure better than a plow except that these organic materials tend to ball up on the tiller's tines. I do suffer along with a tiller, but I'd much prefer a tractor with a plow that could be lifted hydraulically.

Fall plowing is best for my ground, and best for most parts of the country. You leave the land rough-plowed over winter, and it will dry out more quickly in the spring. Smaller gardens can be worked up then with a rotary tiller; a disk is much better, especially on larger plots of ground. I don't mean those flimsy little one-gang disks you can buy for garden tractors. If you want a disk for a larger garden tractor, at least buy one of the two-gang disks. But what I really have in mind as the ideal ground-worker-upper is the large tractor disk that can be raised and lowered hydraulically. You may wonder why I keep stipulating hydraulic lift. The reason is this. With a twenty-five horsepower to sixty-five horsepower farm tractor (the smaller ones) equipped with a plow, disk, and rotary mower that all raise hydraulically, you can operate on small plots of ground or lawn. You don't have to have lots of room to turn around in. You can raise your plow or disk up and back right up to a fence corner, lower the implement and begin to operate.

How fine should you work your soil? Not any finer than you have to. The more you work soil, the more you break down its tilth. We used to disk, harrow, and culti-pack ground three or four times before planting on the farm. Very few farmers do it any longer.

Generally speaking, the finer the seed you will be planting, the finer you ought to work up the seedbed. Bigger seeds like beans and corn don't need as fine a seed bed as carrots. BUT,

In fall, the crop residues should be turned into the soil of your garden. If your plot isn't too big and the residues too bulky, a rotary tiller, shown here, is sufficient to handle the job. An even better technique is disking, which is best accomplished with a farm tractor. Hire someone to do it if you haven't the equipment.

of first consideration, is available moisture. The finer the seedbed, the better it will hold moisture. If you are experiencing dry weather, and the future looks to be holding dry, you need to work soil for planting finer than you would ordinarily. On the other hand, if you are almost sure it is going to rain, and the ground has lots of moisture in it already, you can rough-till and plant and get a good stand.

For example, I just harvested my wheat patch, worked the ground up, and planted soybeans. It was paramount that I got the beans in as quickly as possible, because when you double-crop soybeans behind wheat, you are planting the beans very late. My rotary tiller was NOT doing a good job of working the wheat stubble into the ground, and I had a pretty lumpy mess to contend with. The ground was moist, the sky was dark, threatening more rain, so I went ahead and planted my beans in the lumpy mess. It rained, the raindrops actually firming the seedbed for me, and the beans came up beautifully.

Of all the do's and don't's, one rule is fairly constant for everyone: don't work soil when it is too wet. How do you tell? When soil is right for working, you can just barely squeeze it into a mudball—it will usually crumble apart. Soil too muddy will ball up in your tiller cultivator. It will not turn over crumbly from the plow, but will slither off the moldboard as a wet and shiny slab. If you do work soil when it is too wet, it will probably dry into hard chunks if it is clay ground. Then such ground is just plain miserable until it is soaked by rain and dries out again. Then you need to work it before it does dry out completely to get rid of the hard chunks.

Ground that a good organic gardener has been taking care of is so high in organic matter that you can often work it a little too wet with no adverse effects. The tilth is so good, the ground just won't dry into hard chunks.

For cultivation of weeds you can use a rotary tiller. Or any of the innumerable kinds of garden and field cultivators on the market. If you have sizeable acreage in field crops, you will—must—purchase a row cultivator that fits on your tractor.

Chapter Six

Fruits for the Organic Homestead

Fruits are more difficult than vegetables to grow organically. If bug enemies of fruits are not more numerous, they are at least more persistent and harder to corral. For instance, even if you have only one cherry tree or one apple tree, there is no way you can hand pick the worms or the bugs that lay the eggs that make the worms.

I can't think of a better place in this book to mention a suggestion. Every serious organic homesteader ought to make bugs a hobby and study them. Not only will you find the project fascinating, but you will gain essential knowledge in how to manage a garden of bugs. Understanding the life cycles of bugs is a necessary first step in successful organic gardening. You should know much more about predator bugs (bugs that eat other bugs) than merely that ladybugs eat aphids. The ambush bug eats aphids, too. The tachinid fly, of which there are about fourteen-hundred known species, is another very beneficial bug. It lays its eggs on or in other harmful bugs like the squash bug or the larva of gypsy moths. Another beetle, the Caterpillar Hunter eats both harmful and "good" bugs. Larvae of the Caterpillar Hunter feed on the larvae of the gypsy moth and the tent caterpillar. The larvae of ant lions trap and eat ants. Inside that gob of froth you see on one of your plants hides the spittlebug.

The beauty of bugs in your backyard rivals, if not surpasses, the beauty of birds in your backyard. Exciting things are happening in the world of bugs outside your door—as exciting as the survival of the fittest among animals of the jungle. All you need to watch is a magnifying glass and some time. But the point is that bugs are not all bad, as we humans seem to believe, and are in fact of great benefit to us. I believe that if everyone had studied, or will study, the insect world with as much verve as we study the plant and animal world, we would never have agreed to allow the indiscriminate spraying of bugs with a poison that could kill them all. We "animal lovers" would, with an appreciation of the insect world, have viewed such poisoning in the same way we would view poisoning all the animals in America to get rid of rattlesnakes.

One reason the spraying of mites in northwestern apple orchards has never been quite as effective as growers would like, is that the sprays killed not only the harmful mites, but the predator mites that ate the harmful mites. Several years ago, an entymologist in the State of Washington, very familiar with the problem, told me (after I promised not to quote him by name) that orchardists would probably be better off today if they had never sprayed for mites—at least not until a spray could be concocted that would kill only harmful mites.

Well, enough of preaching to you. I just happen to have started collecting identifying, and studying bugs with my ten-year-old son. I find it more exciting (and easier) than birdwatching or hunting animals, and I merely want to share another good part of the earth with those of you who may not yet have tried it.

A second reason fruit-growing the organic way is harder than vegetable growing is fungal diseases, which generally seem more critical on fruits than vegetables. And there is not, at this time, much organic defense against blights and fungi. About all you can do in humid areas about brown rot is to pray for hot, dry weather. So be prepared for some disappointments. In wet years, you may lose as many as half your sweet cherries, peaches, apricots, plums, and even strawberries.

A third reason why fruits are harder to raise than vegetables is our beloved friends, the birds. Birds, like children, prefer dessert to salad. Here's a few courses of action to help you avoid some of the bird problem.

While birds can be a plague on your fruits and berries, they aren't all bad. Many birds eat bugs that prey on garden plants, so it's good to have some birds, like these martins, around.

Screens

You can frame in and screen a few blueberry plants quite easily, and if you only have about five bushes, you better, or the birds will get them all. If you use the new flexible plastic screening, you will have to build a frame to hold the screening at some distance *away* from the bushes. You cannot just drape the screening over the bushes as some of the advertising might lead you to believe. You cannot drape it over a tree either except a very small one. I've tried. It is one of the most exasperating jobs I know of. A grape arbor might be screened, but you'd have to use framing, and I would wonder if the work would be worth it. Years ago, grape growers like my father-in-law slipped a paper sack over each ripening bunch of grapes and closed the end over the stem with a rubber band. This protected the grapes very effectively, cheaply, and once you got the hang of it, with a minimum amount of time.

Bird-chasers and scarecrows

I think my scarecrow helps a little keeping birds, especially pheasants, away from germinating corn, but not much. I have tried just about every bird chaser known to man, and I can state rather authoritatively if callowly: they don't work. Metal disks bobbing in the wind; dangling, jangling metal strips on a line; hoses draped in trees to resemble snakes; colored banners waving in the breeze. None of it works for me. I think if a person put cats in cages and hung the cages in trees, he might hold the birds off a day or two. The exploding-cannon affairs work if you move them around and don't fire them too regularly, but they are terribly irritating to listen to, and you wouldn't get away with one in a suburban area more than half an hour. I hope.

One trick I haven't tried, but a fellow who has no reason to fib to me says it works—for crows. You simply kill a crow and hang it where you want other crows to stay away from.

I've tried sprinkling rabbit blood around crops I didn't want rabbits chewing on (like fresh young raspberry cane tips), and it seemed to work. I know one gadget that does not work though—those fake owls that are supposed to chase other birds away. Some ingenious people are selling records of the blackbird's whistle of fright or approaching danger. You play the recording over loudspeaker systems and blackbirds are supposed to fly away. Results have not been exactly satisfactory.

Feeding the birds something else

This ploy seems to me the most effective way to control bird damage. There are several ways of going about it.

——1. Keep water around for the birds to drink. Sometimes birds pierce grapes, suck out the juice, but eat nothing. I think they are mostly thirsty. A pan of water is okay, but if you can set a spigot to a slow drip, that's better.

——2. Birds are like humans; they enjoy cherries, strawberries, and blueberries at the beginning of the season much more than toward the end of the season. They seem to tire of a particular taste after gorging for a couple of days. I mention this to save you from a rage that might lead to heart attack when you find the first ripening strawberries with bird pecks in them. I've generally found that by mid-season, the birds are tired of strawberries and are sitting in their favorite perches waiting for the blueberries to ripen.

——3. Birds have a thing about color. Blue, purple and red are their favorites. Yellow and white just don't turn them on. I think this may be a fairly original observation, and you can take advantage of it if you happen to like white grapes, white peaches and yellow sweet cherries.

Here's what I mean. I have three trees of sweet cherries, one a dark red Tartarian. This tree I have given literally to the birds. They like the cherries so well, they eat them before

Color can make a difference. These light-colored grapes have escaped the attention of the birds, who enjoy good grapes as much as anyone. The darker grapes rarely go unnoticed.

the fruit is ripe. (I have a theory that this makes them half sick and that alone deters them from worse depredations of the other trees.) Tree number two is a Queen Anne, a yellowish-reddish variety. The color fools the birds at least until the cherries are ripe, so I get half of them. The third tree is a yellow variety, Yellow Glass, I think. The birds eat only a few of these cherries. And they're just about as good as the others.

Same thing holds for peaches. The birds peck holes into some of the reddish-yellow ones, but the white varieties, they bother less. A juicy white Belle of Georgia is just as good as a juicy Red Haven. So I have both.

In grapes the color difference is quite dramatic. The birds will ruin at least half of my purple Concords if I leave them unprotected, and half of the red Delawares, but the greenish-white Catawbas fool them. I have Concords and Catawbas growing right next to each other, and the latter are rarely harmed.

——4. Years ago, one of the reasons everyone planted a mulberry tree or two around the home was to feed the birds and keep them from feeding on the better fruit. It works, a little. We have many mulberry trees along the road that passes our yard, and occasionally, birds will glut themselves and stay away from garden berries for awhile.

Another aid is to keep your cat under the cherry tree, if you can. That does help. I feed our cats under the blueberry bushes, and the old tomcat will sleep there on hot days. The birds still come in for berries, but they fuss a lot at the cat and don't get nearly as much eating done.

So much for the three big bad B's of fruit growing: Bugs, blight, and birds. Let's consider now the fruits you can raise dependably in order of good to declining dependability.

Apples

The organic homesteader's success comes from taking a wise look at the past together with a searching look at the future, and picking the best of both. The apple tree provides a good chance to use this formula.

Looking back, you find that the apple tree was the first choice of pioneers. The reason? First, the apple proved itself most capable of withstanding attacks of weather, blight, and pest, and still producing a crop. Secondly, the apple came in enough variety from June until cold weather to keep from dulling the taste. Some apples made good jelly, some good pies, some good applesauce, some good fresh-eating. Still others made the best cider, the healthful ubiquitous drink of pioneer times and the source of vinegar. Apples could be dried easily for winter use. And certain varieties would last fresh all winter in common storage.

You can still do as the pioneers did: grow apples without chemicals, and in enough variety to last from summer into the following spring. Either Early Transparent or Lodi Transparent starts you with a green-yellow apple in July, and from there you proceed through varieties like Early Winesap, MacIntosh, York, Rome, Macoun, Golden Delicious, and, for winter storage, Rhode Island Greening.

But with one eye on the past, take a look at where modern apple culture is headed. You can now buy apple trees in about any size model you want: double dwarf, dwarf, semi-dwarf, in varying heights up to standard size trees. Nursery catalogs explain the various rootstocks used and the size tree you get; you have only to pick what you want. For the home garden, and especially for the organic homesteader looking for a steady supply of fruit through the year, dwarfed trees are a bonanza. You get an adequate crop of different varieties in the same space you would get an over-abundance of one or two varieties with standard trees.

Dwarf trees, especially those dwarfed with Malling IX rootstock, often have weak root systems, and need to be staked. That may not necessarily be a disadvantage. Many modern orchardists are growing apples in "hedgerows." You can too, pruning and tying the branches to wires almost as if you were working with grapes. A hedge of apples makes a neat landscaping trick, and you can harvest your apples without a ladder.

I personally like the semi-dwarf trees which grow to varying heights between dwarf and standard. I usually buy my trees on EM (East Malling) VII rootstock. This kind of dwarfing results in a tree about fifteen to eighteen feet tall that does not need staking and produces about the quantity of apples I want.

In shopping the catalogs for my semi-dwarfs, I try always to buy what nurserymen call "spur-type" kinds of semi-dwarfs. "Spur-type" refers to a type of growth where more than the usual number of fruiting spurs appear in a given length of branch. The fruit is thus distributed more evenly and heavily along the branch. The spur-type don't send up

Good nourishment for fruit trees can be provided by cover crops other than alfalfa. Sweet clover is an excellent choice.

many suckers, either. (A sucker is any new green growth that has no buds; it saps the vitality of the tree without producing fruit.) I think you get more fruit from spur-types, but perhaps more importantly, they seem to require less pruning.

Plant young trees the way the directions say you should. A hay mulch around the tree, as far out as the branches reach, will keep the tree alive better during those first critical months after planting. Keep the young tree watered, too, that first year. University experiments twenty years ago proved that a thick hay mulch increased yields on bearing trees, too. Mulch may, however, encourage mice, which damage apple trees in some areas by chewing root and trunk bark. I never have this problem, but a grower I know in Ohio combats mice with a gravel mulch next to the tree trunk. Mice cannot dig through the pebbly gravel—they can't make holes in it!

Be sure not to miss your dormant oil spray in spring. This involves spraying the dormant fruit trees with a miscible oil, which is simply a fine grade of petroleum oil—ninety-seven percent petroleum oil and three percent inert matter. (What is inert matter, I wonder?) The oil coats the tree and actually suffocates any bugs infesting it. Miscible oil is good on scale, probably the best of the available remedies.

Miscible oils are pretty much alike. Lots of companies sell them—I use Ortho's Volck—and they all charge too darn much for them in my humble opinion. For the organicist who will use no other chemical spray on his trees, however, the dormant oil spray is most important.

I strongly advise that you apply it twice: the first time during an early March warm spell, the second right before the buds open. (The oil will harm buds that have already opened.) Do a good job. Make sure the oil hits all sides of the branches.

An organic orchardist I visited in Indiana one summer had developed what looked like a very effective organic spray for his apple trees, applied after petal fall. He used a mixture of powdered rhyania, pyrethrum, lime, and soap. His trees showed no bug damage.

Many of your apples may be knotty, wormy, scabby. But you will get plenty for eating. Blemishes on the apple hurt only its apppearance; a worm now and then can be removed just as easily as removing the seeds.

There is no set way to prune an apple tree. Cut out the dead and diseased stuff. Don't allow weak-angled crotches to form —a branch should grow from the trunk at least at a forty-five degree angle; the wider the angle the better. Beyond that, most pruning is done commercially to ease, enhance, or speed up harvesting. Since harvesting costs are not generally a problem with you, you don't have to take pruning too seriously. Old-timers said to prune an apple tree out enough so a robin can fly through it without hitting his wings on the branches.

Pears

Pears are not as versatile as apples for food, but in my experience they are easier to grow organically. Pears have fewer pest enemies. They are, of course, susceptible to fire blight, but there isn't any very effective cure for that, chemically or otherwise. If you see brown, dead tips of branches on your pear trees, cut them off and burn them.

If you grow two different kinds of pears, you'll get better pollination. Culture for pears is the same as for apples.

It has never been quite clear to me exactly when to pick

Pruning is a good mid-winter project. Be guided by the two reasons for pruning: to keep the tree healthy and to ease harvesting.

pears. The books all say to harvest them before they get ripe. But how much before? I've picked them green, half ripe, and fully ripe, and the results have been mystifying. I've tasted some darn good pears ripe from the tree. And I've picked some green to ripen on the shelf according to the gospel of horticulture, only to find them grainy and not so tasty. But I keep trying.

So far, I can say this much: Let a Bartlett pear (the best all-around variety for my money) get yellow on the tree. But before it falls off of its own accord, pick it very gently so the stem comes off the branch at the right place, not tearing off next year's fruit spur, which is easy to do if the pear is not "ready." Wrap the pears individually in newspaper and allow to stand in a cool place until dead ripe.

However, for picking ripe from the tree, Seckle pears are fine. The little Seckle pears from an old tree like the one in my father-in-law's yard, which has been there no one knows how long, taste wonderful dead ripe from the tree or picked off the ground under it.

Raspberries

Raspberries, red, black, purple, or yellow, are excellent garden fruits for the organic gardener. Of all the fruits I grow, raspberries are by far the easiest to grow organically. I think they perform better without high-powered commercial fertilizers—even by standards of the high-powered commercial fertilizer user. Bugs bother my raspberries very little. Birds even less.

The raspberry's most potent enemy is mosaic, a viral disease. In the last few years, researchers have developed virus-free raspberries, as they did long ago with strawberries. The virus-free plants are available for many varieties, but they cost more. Worth it, too, if you don't have old plants or wild plants nearby to re-infect yours.

Pruning raspberries consists of cutting out dead and weakened canes and topping healthy ones.

Aphids carry the mosaic from infected to clean plants. An aluminum foil mulch, for unknown reasons, seems to scare away aphids if put under raspberry bushes. (Also effective in tests against bugs that bother squash and melons).

Commercial growers plant their virus-free raspberries as far from wild mosaic-bearing plants as possible. At least one thousand feet away, but the farther the better.

Actually all the talk about virus disease in raspberries is a bit much, I think. It is an affectation of garden writers who read too many press releases. Tbe problem exists and is sometimes serious for commercial growers. But I've been eating raspberries grown by people who know nothing at all about virus diseases for forty years. My un-virus-free plants with wild, infected plants a stone's throw away haven't caught mosaic yet. So there.

Once started, a raspberry patch ought to last seven to ten years. After the fifth year, start a new row so it will be in full bearing by the time the old one starts falling off in yield. Work the ground up good before you plant the bushes. I tried sticking them in holes in the sod. Not a very good idea. Half of them died.

Raspberries respond well to mulching. The rotting mulch, especially if it has a little manure in it, is about all the fertilizer raspberries need for a fairly good yield. But the mulch has equal value as a moisture saver. *Raspberries must have an adequate supply of water as the berries are forming.* If you can't irrigate during a drouth at that time, mulch can save the day.

You prune blacks and reds differently. When the new cane —the stalk—of the black (the cane that will fruit *next* year) is about four or five feet tall, nip off the tip. That will encourage side growth, and better, bigger berries. I have two wires strung along the black raspberry row—one on each side about four feet off the ground. This year's fruiting canes are all tied to one side. The new canes are all tied to the other side when they reach five feet tall. This method of trellising keeps the patch from turning into a jungle and makes picking easier. I never allow more than two canes (new) per foot of row. Weaker canes I cut out when I'm nipping tips off the stronger ones. Incidentally, if you want to propagate more plants, don't tip the canes, but let them grow. They will bend down to the ground by August, the tip will push into the ground, and a new plant will grow. Next spring cut off the old cane. Then you can transplant the new plant.

Red raspberries increase by suckering. In spring, you will have to pull out the suckers—any new green canes on the edges of the row—so your row will not spread too far out. You can transplant them, if you want to start a new row, or fill in gaps in the old one. Don't let canes get too thick. Keep the row width narrow, no more than a foot wide, and allow only about four canes per square foot. Prune in winter or early spring to a height of about two and one-half feet. That's for summer bearing varieties. You can prune the same for everbearing or fall bearing varieties, but if you want a big fall crop, prune everbearing canes in winter right down to the ground. You won't get a June crop at all from these plants, but the new canes coming on almost always bear heavier in the fall because none of the vigor of the plants is used up in the June crop.

Purple and yellow raspberries are delicious. I've not found varieties that do well in my garden yet, but I haven't tried too hard yet, either.

Blackberries

Blackberries are a dependable berry for the organic gardener. Treat them like black raspberries. The best way to propagate them is by root cuttings—cut a piece of root into two inch segments and plant.

I was content to go many years without blackberries in the garden because I could pick all we wanted from wild patches. But a couple of years ago, I found a small patch in the woods with berries of a marvelous size and quality. I marked them, came back in the spring, and transplanted a couple of plants to the garden, where they are now increasing.

Strawberries

Though affected by a number of ailments and pests, strawberries make a reasonably good fruit for organicists. Birds can be a problem for some places, particularly at the beginning of the season. Having cherries around will curb the birds' appetites for strawberries. Some diseases can be avoided by growing varieties that are resistant to them. For example, if red stele is a problem, buy strawberries advertised as resistant to it. (Red stele is a viral disease for which there is no cure. It usually announces its presence by turning the root cone red. Red leaves don't necessarily mean your plant has red stele, however; they could mean your plant's getting old or is suffering from a trace-mineral deficiency.) Insect pests have been no problem for me, but rot certainly has. I lose many berries if the weather is rainy during ripening season. One of the reasons is that often my ground is actually too fertile from applications of manure, which causes the plants to become overenthused with nitrogen, grow big and bushy, and produce berries that seem especially prone to rot. Too much nitrogen can also result in fewer berries because the vigor of the plant all goes into leaf growth. I've had that happen, too.

There's no special trick to growing strawberries successfully, but plenty of work. Buy plants from a good nursery and set them out as early in spring as you can. I get an early, a midseason, and a late variety, twenty-five of each, which in the next year will produce between seventy-five and a hundred quarts of berries, depending upon how bad the rot is. Having three varieties ripening at different times not only gives you a longer season, but spreads your risk. A frost might catch one variety blooming and kill some blossoms, but it won't catch all three. Or if early berries get hit hard by rot, later berries might, after a period of dry weather, come along fine.

Plants from a good nursery come with directions on proper planting methods. Follow them. Row widths and spacing between rows vary with the gardener, depending somewhat on how he handles his plants. You can let runners develop into a matted row, or you can prune runners to shape your row any way you want to. I allow four runners to set new plants around each mother plant. The rest of the runners I prune (clip) off as they develop, something that would involve too much labor on a bigger patch. Runners will root by themselves, but I always help mine along by inserting the developing runner through the straw mulch to direct contact with the earth underneath. Gets the runners rooted faster, which means they will develop larger crowns and more berries for the next season. All this hand work takes time, but I get very large berries the next year which means I can pick a quart in a hurry.

After I cultivate a new planting of berries, I mulch the plants heavily with straw (or strawy horse manure as I explained earlier). As the plants grow, most of them will bloom. Pinch off the blossoms, so that the plant's vigor is spent in growing for next year's crop. Actually, you don't have to pinch off blossoms. It's just one of those things a gardener with a small patch can do to get finer berries.

When cold weather arrives, I cover the whole patch with four to six inches of straw to keep the ground from alternate thawing and freezing which heaves plants out of the ground. In the spring after freezing weather has past, I pull this straw to the row middles, where it keeps down weeds and makes a clean bed for the berries to lie on. Be sure to uncover your strawberries before they start growing very much, or you may smother them.

The best time to cover strawberries in late fall seems to be a disputed question. Traditionally, you waited until the ground froze solid, then covered the berries. However, some experts are saying now that the berries ought to be covered before the first hard freeze, since the freeze could injure the plant crowns. I cover mine any time around the first of December that I can get to it.

During the first year of growth, the mulch keeps most weeds out of the patch. Any that make it to sunlight, I pull out by hand. In the second year, the bearing year, no weeding is necessary, since I do not keep the patch a third year.

People ask me why I don't. Weeds are the first reason. To keep a patch weed-free the third year is too difficult. Besides rarely will a patch produce well the third year. You *can* make it produce well, but the time and effort will be better spent on that new planting. Another reason I don't keep strawberries a third year is because of space. With third-year berries, you have three patches to keep up—three areas not producing something else. I think it is more efficient of space to tear up the two-year-old patch after harvest and double-crop that space to fall vegetables.

About varieties, I do not intend to say anything nice about everbearing strawberries. I occasionally read about someone claiming to have good luck with them, but I find it hard to believe. I've tried all of them, I think, and have never had very good results. Oh, you get a few, but in my area, not enough to balance the work involved. Eat strawberries in June. There are plenty of other fruits in the fall.

You should try out a number of the June-bearing varieties. (I should say the spring–summer varieties since in the South strawberries ripen earlier and in the North later. It's a problem trying to include the whole country in one breath. In California, they raise strawberries every month of the year now.) Some grow better in certain regions than in other. Here's a *general* listing of what grows where best, but don't hold me to it.

For Northern states: Midway, Catskill, Surecrop, Sparkle, Fairfax, Fletcher, Midland, Robinson, Premier

West Coast: Northwest, Shasta, Tioga, Fresno, Marshall, Siletz

Florida and Gulf Coast: Florida Ninety
Louisiana and western Gulf Coast: Dabreak, Headliner
Middle South: Tennessee Beauty, Blakemore, Pocahontas, Suwanee
South Atlantic Coast: Albritton, Earlibelle
Northeast and northern East Coast: Same as North; also Jerseybelle, Sunrise, Pocahontas, Guardian, Raritan.

Grapes

Concord, Delaware, Catawba families of grapes (I can't speak for California wine or table grapes) ought to be a part of every organic homestead. The Japanese beetles may chew up the leaves that mildew doesn't get first, and the birds love to peck holes in purple and red varieties, (but not the white ones as I pointed out earlier) but still you will get enough to reward your labors.

Plant one-year-old vines in full sun (in partial shade they're more apt to get mildew) like you would plant a tree, pruning away broken roots and cutting the vine back to two buds. Mulch around the vine, and just let it grow the first year. The second year you can begin to train it to a trellis.

There are any number of ways to prune a grape vine, depending upon what kind of trellis you use. If you are training the vine to the usual two wire trellis, you need to maintain a central trunk and two pairs of side arms. The arms should be about five feet long after pruning. For arms, use branches that are about as thick as a lead pencil. This size bears the most grapes.

I prefer not to give too many instructions on pruning. The only real rule is to make sure your vine does not have more than forty to sixty buds left after you are through pruning in late winter.

As I said earlier, you can put netting over a few grape vines to deter birds, or you can slip small paper sacks over ripening

A properly pruned grape vine has a pair of healthy branches for each trellis wire. After the vine and branches are pruned, tie them to the trellis with binder twine, soft tape, or raffia.

bunches. But if the birds do peck holes in the grapes and suck out the juice, the grape pulp usually remains and can be used for jelly, along with unharmed grapes.

For Japanese beetles on grapes (or anything else) an organic control called Milky Spore disease is supposed to work. I've never tried it. I brush the large irridescent-green beetles into jars in the evening when they are not so fast to fly away. Be surprised how many you can catch in half an hour.

Grapes respond well to high organic matter in the soil. They need potash—a good place to use wood ashes. Seaweed mulch, alfalfa hay mulch, greensand and granite dust are other sources of potash you can use.

Incidentally, grapes are extremely susceptible to weed killer sprays. If your vines lose vigor, or if grapes refuse to ripen correctly, you might check on who's using weed sprays in the neighborhood. Tiny amounts of 2, 4 D carried on the wind, can harm grapes hundreds of feet from where it is sprayed.

Melons

Muskmelons usually wind up being discussed with squashes and cucumbers, to which they are related. And the hazards and how-to of squashes and cucumbers I talked about earlier apply to muskmelons, only more so.

Good muskmelons and watermelons are not easy to grow. In fact, I don't recommend watermelons in the North at all, unless you have a sandy soil that warms up quickly. The small watermelon varieties bred to mature in a short growing season aren't very good. Yes, I know, sometimes they are, but on the whole you are better off sweating over something else. The big, long season watermelons that are good like the South better than the North. I've grown them successfully only one out of three years in my clay ground in Pennsylvania.

So I'll talk about muskmelons mostly, though you can use the same culture for watermelons. I sometimes start melons in peat pots indoors as I do tomatoes. If you get a long cold spring, this gives you a jump on the season so that your melon vines, once in the garden, get the benefit of the longest summer days as they begin to blossom. But unless you have sandy soil that warms up fast, transplanting muskmelons to the garden is fraught with danger.

Transplant to the garden too soon, and the plants just sit there and shrivel up while being attacked by every bug and slug in Christendom. I always put a few seeds directly into the hill with the transplant and four years out of six, the melons I'll eat come from the direct-seeded plants, not the transplants. A word to the wise is sufficient.

I take some pains to prepare the hills. First I dig a hole

where the hill will be, and put a couple of forkfuls of manure in it, topped with a handful of lime. (With watermelons, *don't* include the lime.) Then I cover with dirt and plant above it. Of all the organic practices I know, I believe most firmly in this manure treatment to get *sweet* melons. After the melon plants begin to vine out, I mulch the area around them to keep down weeds and give the melons a clean bed to rest on.

The number of melons you'll get per vine varies. But after eight or so are developing nicely, cut off the tips of the vines. They will be growing out of control, and besides the melons that might set on out at the end of the vines later in the year probably won't develop before frost anyway. Tipping the vines then forces the vigor of the plant into the good melons already developing. They'll get bigger and better.

If the short-tailed shrew is in residence on your place, beware. He likes muskmelons. He will chew a hole in the bottom of a melon, and it will rot away before you know it. Best way I've found to fool the shrew is to cut plastic gallon milk jugs in half *vertically*, punch small holes in the bottom of each half so rain water can drain out, and set the melon gently in the half-jug before it gets ripe. Just lift it gingerly, and slide the container underneath. For some reason, the shrews will rarely bother a melon so protected.

It's easy to tell when a muskmelon is ripe—the stem slips from the melon with an *easy* tug. In fact, it will come off by itself sometimes, but that generally means you've waited a day or two too long. With watermelons, you have to be an expert to judge ripeness. I've tried lots of methods, and none of them work. After you've cut open an unripe one, you can sort of judge how long to wait before cutting the second one open. That's the only way I've been able to master the situation.

Melons fall victim to all kinds of wilt diseases. I know no

remedy. Use wilt resistant varieties. I have the best luck with Burpee Hybrid and Iroquois in muskmelons; resistant Charleston Grey in watermelons.

My manure-buttressed melons never have bug problems *after* the first critical two weeks. You figure it out, I don't know why.

If you concentrate on the fruits already discussed, you will have better results and less frustration than if you go on to others. But if you think happiness is having your own cherries, peaches, and plums, read on.

Cherries

A homestead without a cherry tree seems un-American. If you want to grow one, better grow two—one for the birds. I've already discussed birds and sweet cherries. Birds prefer the sweet to the sour cherry, but in the absence of the former, they will gorge on the latter.

Treat cherry trees like other fruit trees, giving them a good dormant spray before the buds come out. But a dormant spray will not protect sour cherries from the cherry maggot. The small white worms don't really hurt anything, including the person who consumes them. I've eaten many of them, but the idea is distasteful to most people. We seed our sour cherries before they go into pies and jelly, and at that time we discard infected fruit. It is slightly amusing to me to have seen on occasion, people enjoying cherries that were wormy because they didn't know it. The cherry maggot is not easy to see. He looks just like the cherry pulp he feeds on—that and a little protein. Good for you.

Sweet cherry trees are prone to bark injury in very cold winters. You don't find many of them growing where winter temperatures drop much below zero. In wet bearing seasons, brown rot can wipe out sweet cherries and hurt sour cherries, though not as badly.

Otherwise, both sour and sweet cherry trees are fairly trouble free. They go along okay for about fifteen years, then die without warning.

Peaches

The peach is not easy to produce for anyone, and even harder for the organicist. In spring on a warm day before buds open, give the dormant trees a spraying of Volck oil. If San Jose scale has gotten to your trees, give them two oil sprays, and soak them good. Scale doesn't look like a bug but is. Scale insects are almost completely stationary and motionless, and look like—well, like scales. You'll know when you see them.

Two different kinds of borers will attempt to kill your peach trees. Size is about the only difference between them, though the lesser peach tree borer generally concentrates on chewing up the inner bark of branch crotches, while its bigger cousin goes to work inside the lower trunk and even the roots. When you see a small pile of sawdust and gum mixture on the bark of your tree, there's a borer inside. If you can locate the borer's hole, run a wire up into it and smash the varmint. This method is much easier said than done, however. I generally become frustrated, not knowing for sure whether I have made contact with the borer, and slit the bark open with a pen knife. That way the two of us meet face to face, and I always win the encounter. But I sometimes wonder if my knife probings may not harm the tree as much as the borer does.

Peach trees grow better if the ground around them is cultivated or mulched, rather than kept in sod.

In a wet year, again, brown rot can be very bad to peaches. Be sure to clear mummies (dried up peaches) from the ground and off the tree each winter.

Peach trees bloom early in spring when frosts are still probable. Equally hazardous is bud-freezing during the winter. For instance in 1972, you hardly found a peach growing on Ohio and Kentucky trees. Following a wet fall that encouraged late growth, a warm spell in winter caused the fruit buds to swell. Then in one night, the temperature plummeted, freezing and killing the buds.

Still want to grow peaches? I haven't even mentioned half a dozen other diseases you might run into.

On the plus side, I must admit that despite all the problems, we have always had at least a few peaches every year except one. Since it is so difficult to *buy* a good, juicy, tree-ripened peach in many areas, growing your own might be worth the risk for that reason alone. Also, peaches are one of the few fruits that tastes as good frozen as fresh, to me. What we do manage to raise, we earmark for winter desserts.

Plums

Plum culture is about the same as peach culture. I refuse to discuss a fruit I have had no personal experience with, which happens to be the case with plums. I have tried to grow three plum trees so far. All three have died. I intend to plant another next year.

Blueberries

Blueberries require a fairly acid soil. If you do not have a fairly acid soil—pH of 4.5–5.5—forget it. You can make soil more acid with some mulches like oak leaves. But trying to acidify normally neutral soils so blueberries will grow well in them is a waste of time and money.

If you have land that grows blueberries, the fruit makes an excellent organic project. Blueberries dearly love a thick leaf mulch and, at least in my garden, are not bothered by bugs or fungal diseases. (Probably not true in commercial plantings and certainly not true with the wild blueberry which is bothered considerably by the blueberry maggot.)

Five or six mature bushes will give a family of four about all the blueberries they want. But because birds eat so many of them, I advise nine bushes, three early, three mid-season, and three late. Or, six bushes protected by a permanent framed screen. I've not screened my twelve bushes and figure at least half the crop for the birds. We aren't big summer eaters of blueberries anyway. Too many other things to eat at that time of year. Mostly we freeze them for winter. Blueberries, like peaches, preserve excellently when frozen.

Elsewhere I have described how I maintain mulch around the blueberries. Keep the ground moist and cool for blueberries (but not wet). They won't stand much dry weather.

Blueberries are a good crop if you have fairly acid soil, pH 5 to 5.6. Insects aren't much of a bother to blueberries, but weeds are, so a heavy, preferably slightly acid, mulch is beneficial. Peat is another good material for mulching blueberries.

There are other fruits you *can* grow, and perhaps with success. Certainly you will grow citrus where citrus can be grown (birds don't like citrus!), watermelons where they traditionally do well, strictly regional fruits like scuppernongs within their native regions. But I am not going to encourage you to grow nectarines, apricots, boysenberries, dewberries, or loganberries in the North, even though you will read that there are hardy varieties that won't kill back in winter, or even though I know successful nectarine and apricot orchards in the North. Nor will I encourage you to grow quince, currants, or gooseberries, delightful as jellies made from them are. I know from experience that projects with any of the above fruits are not going to pan out for you in the long run. And once you see the way your schedule on your organic homestead fills with work and play, you will understand what I mean when I say you do not have time to fiddle around pulling stems out of gooseberries or wrapping winter coats around boysenberry canes. Our neighbors here in Pennsylvania raised boysenberries, spending an awful lot of time covering the bushes with the onslaught of winter. With due apology to California, the boysenberries really didn't taste that much better than good quality blackberries.

Chapter Seven

The Grains You Should Grow—For You and Your Livestock

The semi-independence of the organic homestead that comes from raising all or most of your food can be more easily attained by careful attention to growing grains, especially corn, wheat, and soybeans. These crops are important to you for at least two reasons:

1.) All three when grown organically are much more dependable sources of food than any fruit and most vegetables.

2.) All three of these grains are useful both as food directly for your family, and, indirectly, as food for your livestock—your source of meat, milk, and eggs.

Corn

For purposes here, there are at least three kinds of corn: field corn for animal feed, sweet corn for humans, and popcorn. Culture is practically the same for all three. There are many varieties of the three types, at least one for almost any and every place in the United States. Corn is king in this country, descended from the native Indian maize. This should tell an organicist something immediately—that corn is acclimated to the normal conditions of this country. Nature

Corn is such an important field crop in the United States, that many farmers plant it year after year in the same fields. Corn is a heavy feeder, and such practices can rapidly deplete a soil. It's better to rotate your crops.

intended it to grow here, and grow here it will. Corn has enemies (most of them imported), but year in and year out, it's the crop farmers in the East and vast Middle West can rely on.

Corn is a heavy feeder on soil nutrients, and will respond well to extra fertilization. It's your best crop after clover or other green manure crop is plowed under. Alfalfa makes the best green manure to plow down before corn because this legume is not only high in nitrogen value, but also in potash, always difficult to get in the soil organically in sufficient amounts. By all means see to it that ground to be planted to corn is sufficiently supplied with rock phosphate. Also lime your corn ground up to a pH of about 6.5. It goes without saying that the more animal manure you get on corn, the better. Corn has been known to handle twenty tons of manure and more per acre without stalk rot or nitrate poisoning. I don't advocate that much—most of us feel lucky if we can get half that amount.

Commercial corn growers on good land often grow "continuous" corn, that is, corn on the same ground year after year. This practice is economically sound, but demands heavy applications of commercial fertilizer and possibly insecticides to guard against pest buildup like Western corn rootworm. It's much saner to rotate corn with other crops. In the garden, I try to stick to a rotation of vegetable, wheat, clover, corn and back to vegetable. On larger acreages of field corn, the organic farmer should follow the time-honored rotation of corn, wheat (or other small grain), clover for two years and then back to corn.

Wireworms, rootworms, corn borers, and earworms attack corn, but rarely are their depredations severe enough to cheat you out of a crop. Rotation of crops is the first control. Plant breeders have developed varieties somewhat resistant to borers, but even when a borer weakens a stalk to the point that it falls over, you can usually save the ear. Earworms make sweet corn look ugly sometimes, but unless you are selling your corn, that isn't much of a problem. Just cut out the part the worm has eaten. On a *small* patch of corn you can control borers and earworms by hand picking. When you see a little pile of sawdust-like material on the stalk right below the tassel, you know there's a worm tunnelling inside—usually upward. Locate the varmint and squash him by squeezing the stalk together. Old-timers used to do this regularly, and I find it to be not nearly as tedious a job as I had imagined—on a small plot.

A drop of mineral oil on the silk of each new forming ear is supposed to keep earworms away—I don't like to advocate a practice I haven't tried myself so I shouldn't really mention that, but others find it effective. I've never found it necessary to fight earworms this way—yet.

Sweet corn

I can use sweet corn many different ways, so it is the only corn I grow at present. What we do not eat fresh, frozen, parched, or ground into cornmeal for table use, I feed to my chickens. They like it better than field corn both in the milky stage and mature. If I were fattening a steer or milking a cow, I think I would still raise only sweet corn, sell the best ears to other people and feed the excess to my one or two animals. In that case I could feed the whole plant either when it was still green or as fodder after the plant had died and turned brown. The stalks could also be cut, tied into bundles and stacked against the outside walls of a chicken coop or other barn building as winter insulation. I followed this practice for several years and kept my chickens snug at zero temperatures in an otherwise drafty building.

But whatever other uses you make of your extra sweet corn, be sure to dry some of it and store it where rodents can't get to it. When you want *good* cornmeal, shell a few ears, grind the kernels in your blender or mill, and presto, you've got it. With a cookbook of whole grain recipes, you've also got the makings of many good, nutritious winter meals.

There are many varities of sweet corn, yellow, white, and crosses between yellow and white. Maturity dates range from around sixty days to one hundred ten. Generally speaking, the longer it takes a variety to mature, the bigger the ear and the bigger the crop.

When should you start planting corn? Folklore says when oak leaves are as big as squirrel ears, which really isn't too bad of a rule if you know how big squirrel ears generally are and have an oak tree close to your garden. A better way, if you want to be very scientific, is to stick a soil thermometer in the ground and see what the temperature is about three inches down. Corn will not germinate worth a darn until soil temperature is up to at least sixty degrees, and will germinate better if the temperature is higher than that.

Once soil temperature is right for corn, you have two alternate schedules to follow to assure yourself of fresh corn until frost. The one schedule is "succession planting," which means you plant a particular variety every two or three weeks from the beginning of the growing season until about the middle of July. The amount planted is geared to the amount the family can eat until the next planting is ready.

This schedule is probably better for those who prefer one variety of corn to all others. The trouble is that succession planting doesn't always work out in reality as it does on paper. Reason is, early planted corn will usually not grow as fast as the seed packet claims, since the claim is based on ideal weather conditions. Late planted corn will mature *faster* than the seed packet claims because the weather is hotter than "normal" and because all late planted crops tend to "hurry" to maturity, as if they knew that in the fight for survival it was urgent that they mature before frost. Anyway, succession planting tends to bunch maturity more than you figure.

The other schedule is to plant all your corn at much closer intervals of time, but with varities of varying maturity dates: for instance, a sixty-five-day corn, a seventy-five, an eighty-five, and a hundred. I favor this method as it gives us a variety of corn flavors. Around the first of May, I plant several rows of a sixty-two day corn, yellow Early Sunglow. Five to seven days later I plant a seventy-eight-day corn, Honey and Cream, which has both yellow and white kernels and is the best tasting corn of all to me. Five more days, I plant an eighty-five-day corn, usually Golden Cross Bantam or a similar kind of yellow corn. In another five days or so, I plant my white corn, usually Silver Queen, a ninety-two-day corn, or perhaps Stowell's Evergreen, which is a one-hundred-day corn. The only other planting I make is during the first week of July when I tear up the old strawberry parch and put it to corn —one of the short season varieties.

Plant corn in rows wide enough to allow for whatever kind of cultivation you intend to use. Tractor cultivators of various makes and models can usually be adjusted to fit row widths of from thirty-six inches to forty-two inches. But if your cultivator will fit between thirty inch rows without cutting into corn roots, then plant to that width. If you are going to weed by hand, by hoe, or by hand-pushed cultivator, you can put your rows even closer—down to twenty inches apart. Narrower rows of corn, if the fertility is there to support that many plants, will generally make higher yields. The corn plants will also shade the ground more quickly, conserving moisture and inhibiting weed growth.

Sweet corn is the type raised for table use, and is one of the few native American vegetables. It remains a favorite, despite the fact that it takes up a lot of garden room for the amount of eating, however good, that it produces. Sweet corn gardeners find a mulch beneficial.

In the row, space the corn plants about six inches apart. Space seeds closer than that if you are planting open-pollinated, older varieties that do not germinate well. If you do not go through your seeds and throw out small kernels that have come from the end or tip of the ear, plant a little thicker anyway. You can always thin if the plants come up too thickly. Professional corn growers plant anywhere from twenty-thousand to thirty-thousand kernels per acre. An organic farmer should stick to plant populations of not more than (and preferably less than) twenty-two-thousand per acre.

Generally, earworms will be most in evidence on your first crop. It often happens that the second or third planting to mature will have little earworm damage. If you plan to sell organic corn, learn which crop is least bothered by the worms in your schedule and make that your biggest main crop for marketing.

On smaller plots of corn, bird and animal damage is worse than insect attacks. Deer, coon, squirrel, groundhog, blackbird, crow, and pheasant all love to see us sweating human beings grow corn.

I also use a scarecrow when the corn is coming up. Laugh if you want, but my scarecrow, moved to a new position every day or so, does fool pheasants into staying away from newly sprouted seed. And we have lots of pheasants. My scarecrow is partially mobile, but easy to make. First drive a stake firmly into the ground (but not too deep, because you will want to move the scarecrow around). Now imagine a clothes hanger hanging from the top of the stake. But instead of a clothes hanger, substitute a thin board half as long as your outstretched arms. By means of a piece of heavy string attached to both ends of the board, hang it from the top of the stake. On this crude "clothes hanger," put an old shirt, the board running through from sleeve to sleeve. Then pin an old pair of pants to the bottom of the shirt. Because the whole thing

A scarecrow won't always keep the crows and other birds out of your garden, but this Logsdon monster does scare pheasants. And what garden is complete without a scarecrow.

hangs loosely from the string over the top of the stake, the slightest breeze will make the "arms" wave back and forth rather realistically. I pin a shiny piece of aluminum foil on the end of each sleeve and put an old hat on top of the stake. I tie the bottoms of the trouser legs together around the bottom of the stake so they don't flap loose. My Frankenstein monster does not scare crows, but he does scare pheasants.

I do not know how to keep raccoons out of corn. I once read that kerosene-soaked rags hung around the patch would keep coons away. Didn't work for me. Some say an electric fence, with the wire no more than six inches off the ground, will do the trick; others say it won't. My sister stakes her dog out by the sweet corn patch during the season, and so far that is the best idea I've seen tried yet.

A little practice will teach you to tell when sweet corn is ready for eating fresh. Some people prefer it a little on the tender side; others a little on the tough side. For cornmeal, allow the ears to dry on the stalk until early winter. If you fear coons will get it, pick the corn as soon as the kernels are dented, strip back the husks, and hang the ears in a protected place until completely dry. Sweet corn kernels shrivel up into wrinkly little things, unlike regular corn, but that makes no difference. For parching over the fireplace on a winter evening or for grinding into meal they are perfect.

After the corn is harvested, you can cut the stalks and tie them into bundles, as I mentioned. We always make a shock and put a pumpkin beside it for a Thanksgiving Day garden decoration. You can also chop up the standing stalks with a rotary mower in the fall, if you don't want to mess with bundles or don't need the shredded stalks for bedding. After chopping up the stalks, it is generally a good idea to fall plow the corn ground. That will destroy some of the earworm cocoons that over-winter in the soil.

Field Corn

Field corn, often called horse corn or hog corn, is not much good to eat as roasting ears, though I have eaten a lot of it that way. Field corn produces much more feed per acre, both in grain and vegetation, than sweet corn. If you intend to grow feed for livestock of any kind, field corn is the most efficient and reliable way to start.

Professional corn growers hit one-hundred-fifty to two-hundred bushels per acre on their yields, even without irrigation. But one-hundred bushels per acre is a good crop, and you should not expect your corn to average any more than seventy-five bushels to the acre.

If you are trying to anticipate how many acres of corn you would need for a given number of livestock, the following figure might come in handy. It takes *about* eleven bushels of corn to put one-hundred pounds of gain on a beef animal. To get a four-hundred-pound steer to one-thousand pounds, figure something like sixty to seventy bushels of corn plus a ton of hay. Good alfalfa hay produces about five tons per acre. So on one acre you could produce all the feed that steer would need. For hogs, figure fifteen bushels of corn to fatten a pig to butchering weight of two hundred pounds. A good milking cow ought to have at least eight to ten pounds of corn in her daily diet. (Figure shelled corn roughtly at fifty-eight pounds per bushel.) A ewe and her lamb together need about three to five bushels of corn a year (along with pasture, hay and supplement of course). About the same for a goat. A laying hen needs about one hundred pounds of feed a year, about three-fifths of which is usually corn. These are all ballpark figures depending upon size of the animal, how much other food it gets, or if you substitute some other grain for part of the corn ration. Nevertheless you can use them to determine roughly how much land you need to raise your own livestock feed.

With a hand-pushed seeder or one that attaches to garden tractors, you can actually plant a couple acres of corn without too much trouble. Acreages much larger than that require a farm-size corn planter. You should be able to find a two-row planter, no longer used on modern farms, sitting back in the weeds behind an implement dealer's lot. The dealer may give it to you. The trick is to be so nice that he or one of his mechanics will tell you how the blame thing works and especially which parts are missing.

As a new farmer, you should by all means make friends with farm machinery dealers. Their used equipment can be your salvation if you have some mechanical aptitude. If not, curry the favor of a neighbor who does. I know that I can make an old corn planter work if I can get seed plates to fit it. But I can't tell you with a typewriter. Reminds me of the ancient corn binder we once bought for $5. It was so old we promptly nicknamed the beast Adam, but we cut and bundled fifteen acres of corn before the machine literally fell apart.

You can actually harvest by hand up to five acres of corn without an impossible amount of work. You have all winter to do it. The best way is to go through the standing, ripe corn and jerk the ears off the stalks row by row. You need a wagon or truck alongside to toss the ears into. If you have a driver, you can move along faster than you might think, once you get the hang of it. In spring (or late fall if you can manage it) after the corn is off, you should cut up the stalks with a stalk chopper or disk.

An even easier way to harvest a stand of corn is to "hog it off," if you happen to be raising hogs. You simply turn the hogs loose and let them eat the corn. This practice, at one time fairly widespread, is somewhat wasteful of corn, but properly managed, the hogs will eventually clean up the grain. Do not, however, turn bovine animals into such a field (or into fresh clover fields). Cows are stupid and will eat until they bloat, a condition that can rapidly bring on death. After the corn is harvested, however, you can turn cows in to clean up stalks, leaves, and any missed grain.

By far the most practical way to harvest over five acres of corn is to pay a farmer to do it with his modern machinery. The cost of custom work varies with crop and region, but is almost always reasonable. In fact, today custom harvesting is often cheaper even for smaller commercial farms.

Wheat

Wheat is another double-duty grain: it can be used as food directly for humans or for animals. Wheat is generally thought of as a crop grown exclusively on large farms. Few gardeners realize that the grain can be grown easily and practically on small garden plots to be ground into flour in a kitchen blender.

There are five general types of wheat. Hard red winter wheat and hard red spring wheat are both used for bread. Soft red winter wheat is best for cake and pastry dough. White wheat is used for both bread and pastries. Durum wheat is used for spaghetti and macaroni. For your purposes, find out what kind of wheat farmers around you grow and do likewise. A farm supply store can guide you in your selection. Try to get certified seed, germination-tested and weed free. If you can't get such seed in an untreated condition, you can use scratch-feed for chickens. The grain won't be certified, and it won't be germination-tested, but it will be untreated, and it will be cheaper than certified seed. Just plant it a little heavier.

I plant soft red winter wheat, the kind usually grown in eastern Pennsylvania. (Winter wheat is planted in the fall, spring wheat in the spring.) I follow the time-honored rule of waiting until after Hessian fly time to plant—after September 15 in this area—to reduce the possibility of damage from this insect.

I rotary-till a fine seed bed for the plot of wheat I grow in the garden. Usually it's a plot that has been in sweet corn. During the last week of September, I sow the wheat, just broadcasting it by hand as evenly as I can on top of the ground. Sow wheat at the rate of five or six pecks per acre.

To cover the seed, I go over the ground lightly with my hand cultivator, or very lightly with the rotary tiller. On a

These are undoubtedly America's staple crops: corn and wheat. Both are double-duty grains, feeding both humans and livestock.

larger area, a harrow is adequate for this operation. Don't worry if you don't get all the seeds covered.

Next I lime the plot. You can apply rock phosphate, a light covering of manure, or other fertilizers at this time, too, or top-dress next spring.

In October, the wheat is green and verdant. Some farmers graze their wheat lightly in late fall. I allow the chickens to feed on mine when there's nothing else green for them to eat —but not too much. The wheat needs plenty of leaves to develop a good strong root system.

After cold weather comes, the wheat turns brown and appears to die. In early March, I sow alfalfa or red clover in the dormant wheat if I am going to need a legume for hay or green manure crop. Choose a quiet morning—too much wind can blow your seed where you don't want it. Also in early spring, the ground has thawed and then frozen slightly again, maybe more than once. This gives you a soil surface perfect for broadcasting seed. If you examine that frozen crust closely, you will see the surface is pockmarked with little pits. The clover seed falls into these little pockets and when the ground thaws again, the mud flows together, giving the seeds a little cover.

The clover will grow only slowly until after you harvest the wheat in summer. Then it grows fast, and will be about knee-high by fall. You can plow it under then or let it go for a hay crop the next year.

When I don't grow clover in my wheat, I plant soybeans immediately after the wheat crop is taken off. They will generally mature before frost. It's one form of double-cropping that works well for me.

Since I have only a small patch of wheat to contend with —one hundred feet long by twenty-five feet wide—I use primitive methods to harvest. When the wheat is almost ripe, I cut it down with a scythe, gather the stalks into bundles, tie the bundles and set them into shocks. On larger acreage, you would allow the wheat to ripen completely as it stands, then combine it. If you have more wheat than a garden patch, but not enough to justify even a small combine, you can usually pay a farmer to combine your wheat for you—which is the practical way to get the job done.

But back to impractical me and my primitive methods. With scythe sharpened, I test the wheat to see if it's ready to cut. The plants should look yellow, with a few streaks of green in the stalks. Chew a few kernels, If they are hard on

Fall is a good time to broadcast rock phosphate, granite dust, and other similar organic fertilizers. They can be spread any time the ground is not frozen.

On the Logsdon homestead, wheat is scythed, then bundled and put in shocks to cure. The wheat is threshed using make shift flails, toy baseball bats are suitable, and winnowed in front of a fan. Primitive maybe, but it works.

the outside but still just a little soft on the inside, the wheat is ready to cut for ripening in a shock. If the grain is milky or quite soft, wait awhile. If the grain is completely ripe—hard and crunchy between your teeth, the plants completely golden-yellow—then the grain is perfect for combining, but too late for the primitive hand methods I use. Mature grain will shatter out on the ground too easily with such methods. That's why you cut it a little green. (The only sure way to test grain for combining is to take a sample to a grain elevator for a moisture test. If moisture content is below thirteen percent, the grain can be safely stored without artificial drying.)

I bunch the cut stalks together and tie enough of them into a bundle to measure about eight inches in diameter at the tie. I then either set the bundles up into shocks (about a dozen bundles make a shock) or I move the bundles directly into the barn for drying if the weatherman expects rain. In about two weeks or less, the grain will be dry enough to thresh out.

Threshing is also a primitive affair. We lay a clean sheet on the sidewalk or patio, put a bundle of wheat on the sheet, then whop the living daylights out of the wheat. The grain shatters easily from the heads. For "flails," we have found that plastic toy baseball bats work best.

I can also thresh out the grain with a rotary mower—at least with a Gravely. I just run the mower blade up over the bundle of wheat easily. A panel of wood propped upon one side of the mower blade stops grain that the blade would normally scatter all over the place, allowing the wheat to gather on the sheet. Crazy, right?

Next, the straw and chaff must be winnowed from the grain. No sweat. We use our big electric fan. The sheet has been brought together like a sack and its contents dumped into a bucket. That bucket is then poured slowly into another bucket right in front of the fan. Light chaff and straw bits are blown away while the heavier grain falls into the bucket below. We usually have to repeat this process several times to get the grain clean. A few hulls stick to the kernels and can't be blown out, so we just leave them. They grind up into the flour just fine.

We store the wheat whole and grind flour only as needed. We store ours in the refrigerator and/or freezer to protect it from weevils. Another way to weevil-proof grain is to put the grain into a shallow pan and heat it for an hour at 135 degrees. (This method works even better for dried beans.) A third way, I've heard, is to put a couple of bay leaves into the container with the wheat (or beans).

We do not thresh out all our wheat for table use. Most of it is fed, a bundle at a time, to the chickens. This practice seems to me to be quite practical for a small flock of hens. If you keep a dozen layers, you would feed them a bundle of wheat every day for their scratch feed, and they could harvest it out of the heads themselves and have the straw left over for bedding. The work of cutting and tying 365 bundles would have been a mere half-day's work for a farm couple of 1850.

The wheat we do use for our own flour my wife grinds in the blender as needed. *Fresh* whole wheat flour imparts a truly delicious flavor to pancakes, hush puppies, bread, and other pastries. Get a cookbook like Stella Standard's *Whole Grain Cookery*, published by Doubleday, and find out what you've been missing. I am almost dumbfounded at times to realize that I was raised with literally acres of wheat at the doorstep, and all those years I believed the idiocy that flour was something you had to buy in stores.

Soybeans

From the standpoint of organics, the soybean offers the most exciting possibilities of all the grains. The soybean is extremely versatile: it is not only food for man and animal, but a vegetable that contains essential proteins we ordinarily get only in high-priced meats. Moreoever, the soybean is also good "food" for the soil—a green manure crop that returns as many soil nutrients, especially nitrogen, as it uses. Soybean plants also make a fair hay, though they are hard to dry. But the bean's main value to the farm is as a source of natural protein for animals, costly when you have to buy it as protein supplement.

Last but not least, the soybean is one crop a canny organic grower might be able to produce in competition with commercial, chemically grown soybeans. The soybean does not respond quickly (if at all) to the stimulation of chemical fertilizers and many champion growers don't apply fertilizer to the crop at all. They rely on fertility built up in the soil over previous years. Secondly, at this writing, there is no insect plaguing the soybean to the point of critical danger. Thirdly, while most commercial soybean growers now use herbicides to control weeds, it is still not only possible, but feasible, to control weeds mechanically.

A couple of summers ago I saw near Tiffin, Ohio, one of the nicest forty-acre fields of soybeans I have ever been privileged to cast my eyes on. It yielded fifty-two bushels per acre—a very respectable yield. It was grown on a farm that has been one hundred percent organic for fifteen years.

My advice to anyone who wants to farm organically *and* commercially: make soybeans your cash crop.

Best way to grow soybeans in the garden or in the field is in rows, a little thicker than you would plant string beans. Plant an inch deep early in the growing season if moisture is adequate; deeper, up to three inches if you are planting in dry summer time, as after wheat in July. If you are growing primarily for animal feed or green manure, use commerical varieties other farmers use in your area. If you want to eat the soybean, too, plant "edible" (they are all edible, but some are more oily than others) varieties recommended for human palates, like Kanrich and Disoy. If you aren't put off by the "oily" taste, you can save yourself a little money by using cheaper (than garden catalog) seed from the farm supply store to produce your table soybeans. The farm store's certified soybean seed, may not be treated, but soybean feed definitely won't be. If you're in doubt, ask. Don't be bashful.

If you have limed and manured the previous crop well, you should be in good shape for the soybean crop. If you are planting after wheat in midsummer on land that was fairly clean of weeds, you could broadcast your soybeans instead of planting in rows. That's what I do—it's easier. Again, I cover the seed by going over the ground lightly with my tiller.

You can treat your seed beans with an inoculant available from most seed stores for use on any legume. The black substance is nitrogen-fixing bacteria which is supposed to help the plant produce more nitrogen to be returned to the soil. Some claim that inoculating increases yields, too. To treat beans, moisten them with water, then sprinkle the black powder over them and stir until all beans have the inoculant sticking to them. Do the treating right before you plant. The bacteria in the inoculant will die quickly if exposed to dry, hot sunlight.

Weed control is critical with soybeans—they won't stand much competition. Make your rows wide enough to accomodate whatever kind of cultivator you plan to use.

When soybeans are ripe, the plants turn brown and all the leaves fall off. Large acreages are harvested by machine, as for other grains. On small patches, you can cut the plants off, put them in a sack, and then beat the sack, or even jump up and down on it. The force will shatter the dry pods and pop the beans out. Then winnow with a large fan, as for wheat.

Soybeans can be fed whole to animals along with other grain, but it's better to grind them into meal. You can make your own soy flour with your blender, but for the animals you'll need a hammermill or one of the new extruders that a few farmers have to grind and cook soybean meal. At any rate, if you do not have much livestock, it will hardly pay you to grind your own feed. Haul it to a feed mill or other custom grinder.

I stress the possibility of double cropping soybeans after wheat if you do not live too far north. I should add that you get better results from soybeans if planted at the regular time—when you plant your first snap beans.

Soybeans offer all sorts of exciting possibilities for the organic homesteader. It returns as many nutrients to the soil as it uses, so it improves your soil while providing you with a protein-rich feed for humans and livestock.

Other grains and field crops are not as important to your homestead as the three discussed above, but a brief mention of a few others may open possibilities of interest to some organicists.

Sorghum

The sorghum family is composed of four more or less distinct members.

——1. Sweet-stemmed varieties grown for silage or for sorghum molasses are sometimes referred to as sorgo. Sorghum molasses, in my opinion, is as good as regular cane molasses and provides homesteaders who don't live in sugar cane country with an alternate sweetener and sugar substitute. Sorghum presses are still operated—I know of two anyway, one in Indiana and one in Tennessee, so I assume there are a good many more. The "sap" from pressing is boiled down into molasses. It's a possibility all homesteaders need to be aware of, anyway. When I farmed in Minnesota, we chewed sorgo stalks like kids with peppermint sticks, and I know it would be no problem rigging up something to squeeze that sweet juice out.

——2. Grain varieties, called grain sorghum, or more commonly, milo. Milo is a good grain to substitute for corn in livestock feed in areas *too dry to grow good corn.*

——3. Broom corn is a kind of sorghum grown exclusively in specialized areas for making brooms. Maybe you want to go into the broom business?

——4. Grass sorghums are grown for hay and pasture. Sudan grass is one type, but some of the new hybrid Sudex-sudan varieties should be familiar to you. These hybrids grow almost unbelievably tall and fast, and thoughtful organicists like Wendell Berry in Kentucky are experimenting with them as an efficient source of mulch, if not feed.

Oats

This grain is not as important as it used to be when it was the main grain fed to work horses. It's still good feed for sheep, and a substitute for other small grains when ground into mash for any livestock. Oats are fairly easy to grow—plant in early spring as soon as the ground is fit, just as you would plant any small grain. I personally feel that growing wheat is better for the organic homesteader, since wheat is more easily converted into human food, and since chickens prefer whole wheat to whole oats.

Rye

I don't recommend growing rye for grain, though it is a dependable cover crop for very early spring and very late fall grazing. For the same reason, rye is often used as a green manure crop even though it is not as valuable to the soil as legumes are. Rye doesn't yield as much grain per acre as wheat or oats or barley. However, organicists will be interested in the fact that rye is *hardiest of all cereal grains and the least injured by disease and insects.*

There are other grains that can be grown on a small scale for both livestock and humans, but I think if you try the ones mentioned, you will have the best results.

Chapter Eight

Livestock for the Organic Homestead

It hardly seems consistent to strive to raise organic fruits and vegetables, and then eat meat, milk, and eggs from animals fed non-organic grains and roughage that might be laced with hormones, antibiotics, and who knows what. For the homesteader, what he can do to remedy this situation is as much a matter of space to grow feed for the livestock as it is to grow the animals themselves. But whether your place is large or small, an acre or fifty acres, there's some kind of livestock project to fit your situation.

While the ideal of a hundred percent organic diet may be beyond the practical grasp of most of us in this regard, nearly anyone who owns any land at all can partially reach his goal. Even on less than a half acre lot, you can keep bees, and bees are very much "livestock" for the small organic homestead. So are fishworms in some ways, and they are compatible with the smallest lot. An acre might support a goat or a few rabbits, or certainly several hens. Two acres can accomodate both a small flock of hens and some rabbits. Another acre makes a few pigs a distinct possibility. Five acres and you can add a cow or beef steer. With twenty acres or more, a flock of sheep becomes a practical spare time project.

A family of four can be expected to eat a minimum of two pounds of meat a day—that's two adults and two children. That comes to 730 pounds a year, about the amount of meat you get from one hog, one steer, and a couple dozen chickens. If you have five acres or more of land, you couldn't do yourself a nicer favor than raising your own meat. After you taste it, I'm sure you'll agree.

Bees

Bees aren't often referred to as farm animals, but that's exactly how you should think of them. And for the amount of time and space you spend on them, bees will return as well as any kind of livestock. Mostly, bees take care of themselves. The whole neighborhood is their "grazing" area. Nor should you feel a bit guilty about that—your bees pay for the honey they collect by providing everyone's garden with better pollination. The honey your bees collect for you is delicious and nutritious and makes you more independent of one of our questionable "necessities" of life—refined sugar.

Fishworms

If you mulch your garden regularly, you will have fishworms as a sideline, without any extra work at all. The earthworm population under my mulch seems unbelievably dense at times. From April until July I could sell a profitable number of worms to fishermen with no more effort than putting an ad in the paper.

For those who want to raise earthworms commercially in beds or boxes, the opportunity to make money seems to be in no way far-fetched. I'm not qualified to speak from experience, but earthworm raisers tell me it is like any other business. The more you learn and the harder you work, the more success you will have. Some insist that profits depend on your location—the closer to a popular fishing resort the better. But at least one fishworm rancher writes that she does most of her selling by mail and has a very good business.

Earthworms are sometimes raised in conjunction with rabbits. The worms live in boxes set underneath the rabbit pens. The manure from the rabbits drops into the worm boxes and the worms feed on it, converting it to valuable compost.

Experiments at the University of Georgia show that raising earthworms commercially in a combination with dairy farming is entirely feasible. Using a regular farm for their tests, researchers piled the manure from the cows into beds and seeded it with worms. No extra care was necessary, though researchers assumed that in case of extended drought, they might have had to sprinkle the beds to keep them moist enough.

The worms converted the manure to compost, which could be sold to nurseries. (Worm-made compost makes excellent potting soil, and weighs less than regular potting soil, thus saving money on plants sent by mail.)

Rabbits convert feed to meat as cheaply as any homestead animal. They also make nice pets, so you must be sure to make a firm distinction between food animals and pets.

The fattened fishworms were then harvested, dried, and sold to cat food manufacturers who have discovered that cats relish the worms.

The cow-manure-fishworm-compost chain provided a nearly perfect recycling system—one that was profitable at each link in the ecological chain.

Rabbits

Rabbits make nice pets, and, in a way, that is the problem. Care for one for awhile, and you haven't the heart to kill it. You have to draw the line very distinctly—especially for your children—between animals for pets and animals for food production. If you don't, particularly in the case of rabbits, you may find yourself very quickly a victim of overpopulation.

Actually, all this latter-day urban sensitivity to slaughtering animals for eating is a bit hypocritical except in the case of the true vegetarian. When I see someone who pretends to be horrified when I talk about butchering chickens enjoying a good steak, I make it a point to remind him that someone had to kill the steer which furnished that steak.

Nor will the slaughtering and butchering of animals make your children callous to "brutality." That's the kind of ridiculous thinking bred by a society too far removed from reality. Some of the most sensitive and gentle men I know and work with were raised on farms where butchering was a weekly affair. They are, if anything, more averse to brutality than the generation now growing up in a purely urban atmosphere. I would bet that if all children could help prepare the meat they eat, we would not now see on our movie screens some of the most loathesome brutality ever visited on the minds of men.

So many books and articles have been written on rabbit care, I won't repeat it all here. In fact, it would be somewhat hypocritical. My last rabbit died.

Rabbits convert feed to meat perhaps more cheaply than

any other animal and that is their main value to you. Moreover, rabbit fryers are good—much better than wild rabbit. And though I believe that if you have the room it is much more practical to raise a pig and a steer, on a small homestead rabbits have a place.

Chickens

While I have never felt that rabbits were practical for most homesteaders, I think chickens are most appropriate. Properly handled, hens provide fresh eggs for breakfast and for cooking, meat for the table occasionally, and your own convenient garbage disposal and compost maker.

I've kept a small flock of hens, (varying from twelve to twenty-four through the year) on our two acres for seven years now without any problems. But even if you can keep only three hens, they can provide you with most of the eggs you need. And in these times when a fresh egg has become a rare delicacy, keeping three hens rewards you far and away more than the effort you put into the project.

The first key to success, and the one most often overlooked, is to make sure your hens have plenty of room. In commercial flocks, a floor plan allows two square feet per layer, but that is much, much less than *you* should allow. My flock fluctuates in number as I bring in new chicks and butcher old hens, but on the average, my hens have ten to fifteen square feet of floor space *each*. That's why I have no odor problems and no disease problems.

Bedding must stay dry, both for hen health and for proper compost making. Not crowding hens insures that bedding will stay dry. Perhaps once a month I add fresh clean bedding of straw, dry grass clippings or old hay. The hens scratch in the bedding all the time and develop a layer of fine compost under the litter. There is never really any manure to haul out. I take out the compost when I need it for the garden.

Vitamin K is produced in the dry, composting bedding, and the chickens, pecking away at bits of this and that in the litter, get enough K so that they rarely have pecking problems common in overcrowded, disinfected coops.

Because my chickens have plenty of room, I have only a single board along three of the walls where they can roost. The droppings fall directly on the ground and the hens in their scratching during the day mix them into the litter, where they are absorbed without unpleasant odors. You cannot handle chickens in this manner even in a normally crowded coop, where a catching board is necessary under the roosts,—a board which almost always smells to high heaven.

The second secret to managing hens correctly on an organic homestead is to feed them without wasting expensive grains you have to buy. In other words, substitute your own feed whenever you can.

Be sure your chickens get plenty of sunshine and room to roam. You'll have healthy chickens, but no odor problem. The proper set-up can ensure the chickens sunshine and space and you a relatively light work-load. They can even be left alone for a weekend stretch.

Chickens will eat almost any table scrap except citrus rinds, In fact, they love scraps. And give them the lettuce, cabbage, and other vegetables from the garden you don't otherwise use up. We also feed ours surplus wheat and sweet corn and get waste lettuce for them free from the grocery during the wintertime. Furthermore, I let the hens outside about an hour before sunset every other night except during snowy winter. This allows them to forage for their own food and supplement their diet with ingredients missing in what we feed them. If you have a larger farm, you should let your chickens have the run of the barnyard most of the time.

All such extra food cuts down on the amount of growing and laying mash you have to buy. The Department of Agriculture estimates that if you have to buy all your feed for twelve laying hens, it will cost you $60 a year. You will also spend seventy-two hours a year caring for those twelves layers. If you grow your own feed, USDA says you will need one acre for twelve layers and forty broilers. Those figures hold up to my experience quite well.

Working at it from my henhouse, twenty hens eat about six pounds of feed a day. Substituting as much home-raised food as possible, I have to buy about $8 worth of feed a month which is about how many dollars worth of eggs we sell. Most of our profit is in getting our own good eggs free—and the compost! In winter, when I have little to feed except purchased grain, the flock is down to twelve. I won't keep hens much beyond two years old.

One thing you should note immediately—there is no commercial profit in a small flock of hens. Don't think you can go out into the country and make a living with chickens. It takes thousands of layers or thousands of broilers to make money, and then not always.

We get twenty-five chicks each year, sometimes all hen chicks, sometimes half roosters. In four to five months the roosters can be butchered and eaten. That meat, if we had to buy it, represents about as much money as we have spent for feed up until that time. Soon after that the pullets begin to lay, so we butcher the old hens. *That* meat about pays for the feed used on the pullets to that point. From then on, the pullet which develops into a consistent-laying hen pays her own way and a little more.

We buy our chicks from the hatchery in June, the "wrong" time according to commercial growers who start theirs early in spring. We wait until June because we have no early egg or meat market we want to hit and so why go to the expense of a brooder stove or heat lamp to keep chicks warm in early spring cold?

Once home, we put the chicks into a big cardboard box or a rabbit pen, leave them under a shady tree during the day and in the garage at night. Right away we feed them bread and milk and a chick starter mash. At this stage, they will repay you every cent you paid for them just in entertainment.

Soon they outgrow their box, and we move them to their own room in the barn, separate from the old hens. When they are nearly full grown the two flocks become one, though I think it would be much better if I kept them completely separated. Older hens may have diseases to pass on. Besides they are terribly mean to the younger chickens when you put them together. For awhile, you have to put several feeders around the coop; otherwise the younger ones won't get a chance to eat. Despite these disadvantages, I've blended the two flocks so far and haven't had trouble.

We raise a chicken good for both meat and eggs. It's a hybrid cross between Rhode Island Red (the best straight strain in my opinion—I like its calm temperament) and White Rock, I believe, called Golden Sex Link. The hens look about like Rhode Island Reds, but the roosters are white. Easy to tell apart when they are chicks.

One more thought on feeding. You can make your own laying mash or grower mash, if you get a hammermill. In rural areas, it should not be hard to find a used one. Grind a mash that is about sixty-five percent corn, twenty percent wheat, ten percent oats, and five percent soybeans, and you'll have a pretty adequate feed if your chickens run loose and if you provide them with oyster shells (which you should anyway).

On such a diet, any meat scraps, dried whey, fish meal, and bonemeal you can get into the feed, is all to the better. If you don't have a hammermill, you can haul your grain to a custom grinder or feed mill and have it ground.

Turkeys

Don't.

Guinea Hens

Good watchdogs. If you have close neighbors, guineas will change them into close enemies with their incessant, godawful noises.

Ducks

If you have a pond, keep a pair. Don't let the young ones start more families. Eat them. Too many ducks make creek and pond banks unsightly.

Geese

I have not raised geese, so cannot comment, but one trait of this feathered friend should interest you.

Geese are good weedkillers. Penned into a strawberry field and fed only a small amount of grain to keep them hungry, the geese will forage up and down the rows and eat the weeds but not the strawberry plants. I have seen geese used commercially to weed mint fields, and they were doing a fine job of that, too. Organic, commercial strawberry growers may well want to try this old practice again, as a way to avoid using herbicides.

Goats

Goats are especially useful for people allergic to cow's milk. Or for people who can't be happy without their own source of milk but who do not have enough space to keep a cow. Other than those two situations, I don't advise goats—I think cows are better. Goats have always been associated with organics more than cows, a phenomenon I have never been able to understand. The whole experience of agriculture in milk production has been toward cows and away from goats, but I suppose it does little good to argue. Those who like goats will go on raising them, and those who like cows will stick with them. At any rate, other sources of organic information have covered goat raising thoroughly and much better than I could do. I'll talk about cows.

Cows

The right cow—by which I mean a good one—is a four-legged cornucopia for any organic homesteader who appreciates good food enough to milk her twice a day. The milking is very much the rub when you make a decision to keep a cow—or a goat for that matter. A cow must be milked every day—twice every day—Sundays, holidays, during vacations, without exception. It doesn't take long to milk a cow, even by hand, but someone has to be there to do it.

The only effective way out of this dilemma that I know of is to find another homesteader in your area who has, or wants to keep, a cow. Make a deal with him: you will milk his cow during his vacation or during some emergency, and he will do likewise for you. This is the way farmers have done for years—probably centuries—and there's no reason why it won't work for a couple more centuries.

But I would still hesitate to buy a cow if I were the only

You can milk inside or out, but once a pattern is established, stick to it. An animal used to being fed while being milked will balk at having it any other way.

member of my family who could milk it. Twice-a-day milking of even one cow gets to be old in a hurry, and if your wife (or husband) states in writing he will take his turn, or if you can convince a teenage son that milking a cow by hand is the best way in the world for developing strong wrists for baseball and basketball, then keeping a cow will not become a burden.

If you decide to keep two cows so that when one is dry, the other is giving milk, you might investigate the purchase of a milking machine. Some systems originally made for small herds are obsolete now, and you can buy them very reasonably. Check your area's milking machine dealer—Surge is the largest manufacturer and has a representative in most dairy farming communities—or farm auctions or just ask around.

Once you have solved the milking problem, you have guaranteed your family all the milk, cream, butter, cheese, cottage cheese, and ice cream they want. And the cow's offspring can be butchered for your beef supply.

The Holstein cow is the most popular breed among American dairymen because on the average it gives more milk than other breeds. Since there are more Holsteins around than anything else, you have a better chance of finding a good one.

Holsteins are often flighty, and they can give more milk than you need. A Guernsey gives less, but their milk is richer in butterfat. That's better for you, because you want butter and cream, too. Jerseys give even richer milk than Guernseys.

The disadvantage of Jerseys and Guernseys for organic homesteaders is that their calves are small and do not make much meat for butchering.

In my opinion, a good Brown Swiss cow is the best selection for someone trying to raise his own food. Brown Swiss are big animals, and their calves can be fed out to good beef. Brown Swiss are very calm cows, easy to handle despite their size. A good one gives plenty of milk, and it is usually richer than Holstein milk. But if I couldn't get a good Brown Swiss, I'd select a Holstein—one with a record of giving richer than average milk. Average is a butterfat content of 3.5 percent. Jerseys have been know to hit five percent and more. A Brown Swiss should be around four percent and possibly higher. A few Holsteins give four percent milk. That's the kind you want.

A good Milking Shorthorn cow also gives rich milk and beefy calves for meat, but this breed is hard to find anymore.

How do you find a good cow? The best way I know of is to make friends, or at least get acquainted, with a good dairy farmer. Tell him exactly what you have in mind and grin a little bit. I used to be a dairyman, and I know how my brethren think. They will help you, either by selling you a cow themselves or by directing you to someone honest who has cows to sell. Dairymen will help you because, whether they say it out loud or not, they will admire your pluck (or insanity, take your choice) in deciding to produce your own milk.

I don't advise buying a cow at an auction unless you know the herd she comes from, and/or the owner. You can make use of the services of a dealer who buys and sells cows for a living. As in any other business, there are trustworthy dealers and some not so trustworthy.

The price of a cow depends on many things—there are Cadillacs and Volkswagens in the cow world, too. A purebred Holstein with an annual production average of twenty thousand pounds of milk might cost you $2,000 if you can buy her at all. Uncle Wilber's gentle, 15-year-old Bossie with a bag that drags the ground you might be able to buy for $100. You want neither the best nor the worst, and should expect to pay, in 1973, around $500. You might find a fair to middling cow for your purposes for $300. In some areas on some days, that same quality cow might go for $600. It's a lot like buying used cars.

I believe the best path for the newcomer to take in the cow business is to buy a weaned heifer calf of a fairly good cow and raise her yourself. You'll pay around $60 for her unless she comes out of real pedigreed parents, in which case you may pay a lot more. As you raise her, the two of you sort of grow and learn together about what makes cows tick.

If you have a barn on your homestead built in the days when barns were barns, it will contain a cow stable or stanchions of some kind. Get your heifer or cow used to being tied in her stanchion as soon as possible: the younger you start her, the better. It is possible to train a cow to stand for milking without tieing if the idea of walking out to the field with your

If you have a barn, get your cow used to being tied in a stanchion as soon as possible. If you have only one cow, hand milking is best, but with several, a milking machine can be beneficial.

bucket and milking the cow where she grazes appeals to you.

Better though, is to construct a stanchion if you don't have one. Or buy some at a farm sale. A very cheap but practical stanchion can be made with two pieces of two-by-four. One piece is fastened solid in front of the cow's grain box, the other held at the bottom by a bolt in an oversized hole. The cow sticks her head between the two boards and by means of a rope, you pull the movable board toward the solid one at the top, thereby clamping the cow securely in her "stanchion."

In handling a cow, the only rule is to establish a routine and then stick to it. Ideally, cows should be milked every twelve hours, but if you follow a schedule of milking at 7:00 A.M. and 5:00 P.M., the cow will adapt to it. If you start out milking from the right side of the cow, she will expect you to continue to milk from the right side. If you feed her grain while you milk her, she will object if you try to get away with milking her without grain feeding.

Either take your cow to a bull on a regular farm for breeding, or have her bred artificially. To keep a bull would be the height of folly.

How do you know when your cow is ready to breed? Usually, you'll know okay. She will start acting nervous, flighty, stubborn—not at all her usual agreeable calm manner. Often she will not give down her usual amount of milk. If you have another cow around, she will try to ride the one in heat—or vice-versa. A mucous-like discharge from her vagina may be visible. Don't be surprised if your cow does not get pregnant the first time the vet artificially inseminates her. A cow often does not at the first natural service either.

Gestation period for a cow is about the same as for humans, nine months. About four to six weeks before the cow calves (gives birth), quit milking her. She will no doubt be giving only a small amount of milk anyway—if any. You can taper off, if she is still giving an appreciable amount of milk, by milking her just once a day. Once you have her "dried up" you have three or four weeks of freedom from milking (unless you have two cows). Make the most of it. If you have been planning a trip, now is the time to go.

A cow will generally have her calf without any help from anyone, but occasionally there may be complications. In normal delivery, the calf's head and front feet emerge first. If the calf is coming any other way, call your veterinarian. Perhaps the second time, after you've watched once, you can handle difficult births yourself.

The usual problem is when a calf's head, or more often its front shoulders, are too large or turned slightly. If you decide to play midwife, first make sure the calf isn't suffocating with mucous up its nose. Then take hold of the front feet and pull. Don't pull straight out, but slightly downward toward the cow's bag. Twist the calf *gently* one way and then the other. Usually a few hard tugs and the calf will slide right on out. If not, better call the vet.

The cow will lick the calf more or less dry. It may even eat some of the afterbirth, so don't be alarmed. With normal births, the cow will deliver the afterbirth cleanly within half an hour. Sometimes a vet will have to clean her out.

Let the calf suck the cow at will for the first three days. The cow's milk during that time is called colostrum; it's necessary for the calf but not fit for human consumption. At the end of the three days you can start milking and saving the milk for the table. In a commercial herd, the calves are often put on milk replacer at this time so the cow can go unhindered into the milking line again. In your situation, continue to let the calf drink from its mother. Let it have what it wants for while, and you milk the rest. After a few days, start penning the calf away from the cow, allowing them together twice or three times a day for nursing. While the calf nurses, you milk. Let the calf have two quarters; you take the other two.

You can veal your calf at about eight weeks of age or wean it and feed it out either to butcher or to raise for a replacement cow. If you feed bull calves as you would a fattening steer, they produce good meat if slaughtered before the animal is a year old. You don't *have* to castrate bulls. Older bulls make very good hamburger. Whole-bull hamburger, as we used to call it, is excellent, especially if you like your hamburger a little lean.

Meanwhile, your cow has been eating from freshening some eighteen pounds of ground grain a day plus all the hay or pasture she wants. If everything goes right, she should increase her milk production for awhile, and you should give her a little more grain as production goes up. Generally speaking, grain should be fed on the basis of one pound for every three or four pounds of milk. Depends a lot on what else you are feeding. On a very good quality hay, or a corn silage with lots of grain in it, you might not have to feed as much grain.

Weigh your cow's milk every other day or so. When her milk output levels off, stop increasing her grain ration. As she continues through her lactation, she will start to decrease production after perhaps five months. Decrease grain feeding at the same time, a pound of grain less for every three pound drop in daily milk output.

While a cow can eat all the hay she wants, or grass pasture, you must be careful not to overfeed grains or fresh legumes. As I mentioned before, a cow has no sense in this regard and will go on eating fresh clover or corn, if she has access to it, until she bloats. Severe bloating will kill her if you or your vet arrive on the scene too late. Especially do not let your cow into a field of fresh luxuriant clover or alfalfa. A mixed legume-grass pasture, however, rarely bloats a cow.

After your cow has calved, re-breed her the second time she comes in heat, not the first.

Keeping a cow is no project for the lazy, and I have deliberately made it sound more complicated and troublesome than it is. I doubt if keeping a cow is for everyone.

However, the rewards are often worth it, and the job does not have to be overly difficult. I just visited a farm in southern Indiana where a small herd of Jerseys live that are not milked regularly at all! Impossible? No. The Jerseys are bred to Herefore bulls and the offspring run with their mothers like beef calves. If the cow has more milk than one calf needs, the homesteading couple who own the herd will buy another calf to put on her. When they want milk for themselves, it's available. But they are not tied down to the necessity of milking every day. And they figure the sale of their calves is at least a break-even operation.

Beef

If you keep a dairy cow and raise its calf for beef, you produce that meat much more cheaply than if you raise a

Raise a steer for your own beef. If you haven't a barn, you can buy a "feeder" and fatten him up over the summer, eliminating the need for shelter.

non-dairy beef animal. But if you can't or won't keep a milk cow, yet want your own beef, you'll have to go the more expensive route. If you have your own feed you will still save a little money.

You can keep a beef cow solely for the purpose of raising a calf for you to butcher, but I would advise, instead, buying an Angus or Hereford "feeder," as they are called. A feeder has been running with its mother all summer. You buy it weaned, vaccinated, degrubbed, and dewormed. It will weigh about four-hundred pounds and you'll probably pay around $150 for it, depending on the market.

For the small organic farmstead that has pasture but is short on barn space, buying a steer in the spring to fatten through the summer is a better use of your resources. After the pasture season, the animal will probably weigh around eight-hundred pounds. You can then finish it on grain up to one thousand pounds or more, which makes the meat even better. However (I will say this at the risk of having beef men laugh at me), if your animal has had good grass pasture and weighs only 850 pounds, you *can* butcher it without further fattening, and the meat will be very tasty. I guarantee you it

Hogs have a bad reputation, but if you have a liking for pork, ham, bacon, and sausage, by all means get one. If you care for the hog properly, you'll find the reputation is ill-deserved.

will taste better than what you buy in the store. I have never been able to figure out what the supermarket or meat packer does to meat, but it has a harsher taste than what comes directly off your farm.

Raising your own beef is very simple. It requires only a strong fence. Butchering is not so simple but odds are in your favor that you can find someone who will show you how to do it if you offer him some steaks. But by far the simplest way to get past the butchering chore is to hire the guy who runs your local meat locker. That's what he's in business for.

Hogs

Farmers traditionally referred to the pig as his "mortgage lifter," by which he meant that hogs were the most dependable money he could rely on year in and year out. Hogs knew how to care for themselves in the old days and required little attention. They foraged for roots and acorns in the woods, kept snakes out of creeks, and cleaned up the corn that cattle did not digest but passed on in their droppings.

Raising hogs commercially today is a far cry from the old "root, hog, or die" philosophy, but you can take a cue from the past and raise a pig quite inexpensively.

First, make sure of your zoning regulations. People who compose zoning ordinances are not known for either ingenuity or imagination. Someone has told them that five-hundred hogs squeezed into an acre lot stink to high heaven, so they deduce that one hog on two acres must stink a lot, too, which is a shame. A hog is cleaner than a dog. In fact, a hog is naturally housebroken unless you crowd him in with so many of his brothers and sisters that he forgets his manners.

The hog's bad name comes from the way humans have handled him. If you put five-hundred cats in the same space that five-hundred commercial hogs are forced to occupy, the odor would be much worse than anything that ever emanates from a hog barn. Because the hog can survive under conditions many other animals would not tolerate, man takes advantage of him.

If you are free of zoning board nuisances, or if enlightenment has come to your board, you can keep a hog in a pen in your backyard as easily as you can keep a large dog (easier), if you give him enough room. Enough room is two-hundred square feet, in my opinion, kept well bedded. Your hog will use one corner of the pen for a bathroom if my experiences with hogs holds true. You should clean this corner out once a week or more frequently. The hog will keep the rest of the pen fairly clean. I used to fix up a six-inch-high partition in one corner with a couple of two-by-six's and fill the area with sawdust. The pig would readily use the sawdust platform for its litter box. The sawdust absorbed the urine and a couple of quick scoops with a shovel every day kept the area clean and added manure to the compost or garden.

On small homesteads, don't mess around with a sow, boar, and litter of pigs. Buy a baby pig, just weaned, either a female or castrated male. What you are buying is what farmers call a feeder pig. It may cost you anywhere from $10 to $20 or more depending on market prices, but that way you can raise enough pork and lard to last you a year without feeding and housing momma and poppa pig the whole year.

My sister raises a hog every year on her two-acre-place. She has kept close records of cost and figures she does not save much money if she counts her time as labor. By the time she has the hog butchered at a local locker plant, she says, she might just as well have bought the meat from the locker. However, she finds her own meat "far superior to what I used to buy," and even her husband, who thinks raising a pig is a bit zany and so is not quite as biased in favor of the home variety, agrees that the meat does taste better. "Not only that," adds sister, "but I get my lard more or less free."

And since she gives some lard to me every year, I can vouch that when my wife uses it for cooking, as in pie crust, it does make a difference.

A hog likes to eat grain, good hay, pasture, milk, meat scraps, acorns, fish, table garbage, and just about anything else you can think of. But for the best meat, they should be fed grain, preferably corn.

It takes about twelve to fifteen bushels of corn to fatten a feeder pig to butchering weight of two-hundred pounds. If you are mixing your own feed, you will want to grind in some soybeans for protein and perhaps subsitute oats or wheat for some of the corn. Feed what you have on hand. You don't have to follow commercial feeding rations or schedules. In fact, a hog fattened more slowly than is usual in a commercial lot may taste just as good, especially the hams and sausage if you intend to smoke them.

If you have more than a couple of acres, or a regular small farm, you can let your pig graze pasture. Hogs will root up the ground in search of grubs. Farmers put rings in their noses to stop rooting, a rather unpleasant job. The best way I ever raised a hog was in an oak woodlot through which ran a never-failing stream. The hog could root in the lot all it wanted because there was no sod there to ruin. Water was available for drinking and for wallowing in hot weather, and the oak trees shed a good crop of acorns, a nut hogs relish. I fed it ear corn, choice alfalfa hay, and table scraps, and it made very nice pork indeed.

A hog is very skilled at getting through fences. If yours does, put it down to one of the following: 1.) the fence is in poor shape; 2.) you aren't feeding the hog enough to keep it satisfied; 3.) you have made a pet out of the hog, and it is lonely for your company.

A pig turned into a pet wants to follow you around. Sometimes you find him squealing at the kitchen door when you

With careful management, a hog can be fenced into a woodlot. He'll root for grubs and eat the fruit of the trees.

get up in the morning. As a pet, a pig can be rather lovable and very difficult to kill for meat. You'll look pretty ridiculous though, keeping a huge hog around until it dies of old age.

The only hard part of raising a pig is butchering it, which is why I advise you to do like my sister does and hire someone for the job. However, if there are other homesteaders who raise hogs in your neighborhood, why not get together, share your knowledge and your muscles, and enjoy an old-fashioned Butchering Day. You just might find yourself host or hostess to one of the most successful parties you've ever thrown. Be sure one of you does know something about butchering. Or find a retired farmer (or better yet, one not so retired) to direct the celebration. If you dangle a bait of pork tenderloin, and perhaps some good bourbon, your search will be much easier than you thought. Once you learn the ropes, butchering day can become an annual early winter affair that you and your children will cherish as long as memory lasts.

Sheep

Whether you want to keep a couple of sheep on a small, practical homestead depends upon how much you like to eat lamb. (Fresh lamb is delicious, but many people have tasted only old mutton and are prejudiced against all mutton.) Some small homesteaders keep a few sheep only to keep down weeds and grass in lots too large for handy mowing.

Personally, I doubt if it's practical to keep sheep on a small homestead. However, the animal is just the ticket on a larger farm that contains mostly land too rough to cultivate. Many homesteaders new to the country do buy such tracts of land because they are cheaper—more space for the money. A fifty-acre farm of this kind, especially if there's an old barn on it, can be utilized practically by a small flock of sheep. Sheep will eat weeds and rough grasses better than cattle, and keep your place looking fairly neat without much help from you and your mower. Managed correctly, you might even make some money from the sale of lambs.

Such a project is especially suitable for homesteaders who have a job elsewhere, as I assume you have. Sheep take very little time except during lambing and that comes in late winter–early spring when you have nothing much else to do around the place anyway.

Expect to spend no more than five to seven hours of labor per year per ewe. If you want to know how that compares with other animals, somebody has statistically determined that the labor requirements for a one hundred-ewe flock are about the same as for five dairy cows, twenty-five beef cows, thirty-five feeder cattle, or eighteen litters of pigs.

If you decide to try a small flock of sheep, start with five or ten ewes and a ram. Learn as you go along. Save out promising young ewe lambs until you work up to a flock of forty ewes, or until you have a flock equal to the amount of pasture you own. On rougher land don't figure on pasturing any more than three mature animals per acre.

A mature ram can service about forty ewes. Keep him separate from the flock except during breeding season—September–October if you want early lambs coming in February–March, or November–December for April and May lambs. Early lambs sell on a higher market, but I advise you for reasons I'll give later to schedule late lambs, since you are not commercially dependent upon squeezing every penny of profit possible from your small flock. At any rate, the breeding season should last no more than fifty days.

The flock can stay on pasture the whole year, even in winter when you can feed hay and cornstalks right on top of the snow. The only time the ewes need to be inside is during lambing. A ewe should have a small pen of her own while she lambs and for a couple of days afterwards. Such pens can easily be made by hinging together two eight-foot-long and four-foot-high gates. Set them in the form of a "V" with the open end against the wall, and you have an instant pen.

After a few days, the ewe and her lamb can be turned back into the flock. Lambs and mothers always seem to be able to find each other. If one ewe loses her lamb and another ewe has twins, you can sometimes persuade the former to adopt one of the twins. It's a good idea, too, especially if the second ewe does not seem to have ample milk for both lambs.

To fool a ewe into adopting an orphan lamb after her own lamb has died, my father used to skin the dead lamb and drape the skin over the orphan lamb. The ewe, who distinguishes her own by smell more than by sight, would be conned into allowing the orphan to nurse and in a day or so would adopt the new lamb.

Most sheepmen cut the tales off their lambs (called docking) and castrate male lambs at two weeks of age. In a small flock, you may not *have* to dock the lambs. It's done because usually a sheep's tail will become so coated with manure as to result in health problems. But I've seen some healthy yearlings (not in commercial flocks) with their tails still intact.

Ewes need about five pounds of food per day during lactation; four pounds per day the rest of the year. A hundred ewes and their lambs need about three-hundred bushels of corn or other grain equivalent per year, fifteen hundred pounds of protein supplement and twenty-five tons of hay in addition to pasture. The supplement can be supplied in the form of soybean meal, as previously mentioned. Always remember that if you have good quality legume hay, you can cut down on the amount of expensive protein supplement you might have to buy. If you ask your county agent, he can tell you where to get your hay analyzed for protein content.

Lambs are cute but not necessarily practical for the small homestead. If you have the space, however, a flock of sheep won't take much of your time.

If lambs are born in April–May, they can very soon go right out on pasture with their mothers.

Pasture is the cheapest way to put weight on lambs, and late lambs take best advantage of it. When three months old, wean the lambs, and put them on the best pasture available, if you have any left (or intentionally saved some for this purpose). Or put them on good hay and some grain. When they weigh ninety to one-hundred pounds sell them.

You can shear lambs when you wean them. Rams are sheared right before breeding; ewes usually in late winter up to two weeks before lambing. The wool is worth money—sometimes it covers the biggest part of the feed bill.

When my grandfather weaned his lambs, it was usually August. Not much good pasture left. He turned the lambs into his cornfields. They ate the lower leaves off the corn without harming stalk or ear. (Experts say that if you let your lambs graze corn this way you should vaccinate them for enterotoxemia, whatever that is, but I bet Grandad never did.) After the lambs were through, he turned his hogs into the corn field, and the hogs ate up most of the ears of corn. Then in late fall, he turned steers in to clean up leaves, stalks, and corn the hogs missed. The animals harvested his crop. Grandad may not have been the most industrious farmer around, but I think he was pretty smart.

Chapter Nine

Help from the Wild

You don't have to know as much as Euell Gibbons (though that would be nice) to make use of nature's free gifts. Running your homestead properly, you won't have time to gather all the wild food that's out there anyway. Concentrate on a few wild things every month of the year—those which are easiest to collect or return you the most for your labor. Here's a sort of yearly schedule of how the Logsdons (all a bit wild-eyed anyway) go Hunting and Gathering For Fun and Profit.

January: Wood-cutting time and it applies to you whether you own wooded acreage, a hedge growing up in thickets, or simply too may trees on your lot (the latter a condition that almost always befalls a suburban house about twenty years after it is built and landscaped). In the last two cases, make your work pay double. Cut up wood into fireplace or stove-length logs. Make poles, posts, and trellis material out of what's left over. Burn the brush on your garden plot so the ashes help your soil. Remember that a garden demands an almost endless number of poles and stakes, and if you have to buy them, the cost becomes considerable.

On a larger acreage you should have a woodlot. Think of it as your storehouse of barn building material, feed-racks, fence posts, gates, pens, stanchions, siding, rafters, tool handles, maybe even a cabin or house. Lumber, even for the most casual of purposes, is now high-priced, and your woodlot can be profitable as well as pleasurable.

Here's the homesteader's store of fence posts, building material, and firewood. A properly managed woodlot can be a profitable as well as pleasurable part of your homestead.

First make an inventory of the wood you have in standing trees. If you have signed your farm up with the Soil Conservation Service, the conservationist will make sure you have the services of a forester, when you ask, to cruise your timber and evaluate it.

In forest management, the old farmers used to divide their woodland into three equal parts, and each winter worked on one part, rotating to another the next year. That's still a good method. On the one-third you work on this winter, you thin the stand, allowing better varieties room to grow and cutting out worthless kinds. You clean up dead wood for firewood and harvest mature trees. The logs go to the sawmill to be cut into the lumber you want. The larger branches are cut up into firewood either to sell or to use yourself.

Many people do not like to cut down their trees. That's fine. But don't feel guilty if you do cut your *mature* trees. A tree is like any other plant: eventually it will start to die out in the top and sooner or later rot out inside. To allow a good white oak to die and rot of old age is not very sensible. Just make sure when you cut one tree down that another comes up to take its place. In a properly managed woods you won't have to plant another tree. There should be a little one close by already growing. If you graze your woodlot however, no young trees will survive to take the old ones' places. That's why you don't use your woodlot for a pasture if you are managing it properly.

Some farmers eat their cake and have it too in this regard. They will fence off an area of an acre or less in the woods and not allow cattle or sheep inside the fence. Trees then begin to grow within the fence and in a couple of years are tall enough so grazing animals can't harm them. The farmer than moves the fence to another area and repeats the process. They renew the woodlot and graze it at the same time.

Hardwoods are more valuable, of course, than softwood

trees. Black walnut, black cherry, white oak, white ash, maple, and some hickories are all relatively valuable today, especially the first two. Walnut and cherry are so valuable, in fact, that thieves are stealing them right out of the woods.

Different woods have different uses for building. White oak and elm are best for sills. Ash, maple, beach, elm, oak, hackberry, and sycamores can be used for studding, sheathing, roof boards, plates, joists, and rafters—and also for flooring. Hickory is excellent for flooring. Basswoods can be used for sheathing, plates, rafters, and joists. Yellow poplar (tulip tree) makes good sheathing. Don't overlook cheap old cottonwood—it makes passable studding. Green white oak won't rot in water for years and years, if ever. That's why it was used extensively for water mill-wheels. White pine is similarly rot-resistant. But redwood and cypress are the two woods most commonly used for wooden vessels that will contain water or moist material like silage. All the old wooden water troughs in the part of Ohio where I grew up were made of cypress. Ash is a good strong wood where resilience is needed—it was the wood most often used for wagon tongues. If you are faced with the task of putting a new tongue on an old piece of farm machinery—and you just may be—ash is the wood to use. You can use ash, and hickory, too, for pitchfork and shovel handles. Hickory can be bent fairly easily. Just soak the hickory sticks a couple of days, bend to shape you want, clamp in that shape, and let dry. Beech is a hard wood and is used where wood takes a beating—as in a threshold of a doorway.

If you really get serious about subsistence living, you might like to know that ash wood makes good baskets. Take the bark off a small ash and soak the log. Once soft, beat the living daylights out of it. If you experiment a little, you'll find that in a little while you can start peeling off long strips in layers. The strips you braid together into the basket of your choice. I am not a basket maker, and I should quit before I get in over my head. But I'll add that stripped willow shoots make baskets, too. And oak splines. Making baskets by hand from scratch is still a craft very much alive in England. In fact, a great many country crafts that use material free for the taking are alive in rural England. A good place for modern homesteaders to visit.

Use a chain saw when you cut down trees. The old crosscut saw, except in the hands of experts, is something akin to torture. So is a chain saw after six hours of use. When you saw your logs into the length of board you want, be sure to leave about three inches more on each end. In other words, if you want the sawmill to cut out sixteen-foot boards, be sure the log is sixteen-and-a-half feet long.

After a log is cut into lumber, the boards must be stacked properly to season. The whole stack should be two and a half feet off the ground. Then stack boards in layers with inch sticks between layers and an inch between each board for proper ventilation. The layers of boards should slope slightly —at least one inch to the foot—so that water will drain off fast. Softwoods will dry in a couple of months; hardwoods in four to six months.

There are so many uses for woody, fibrous materials. English roofers in rural areas can still put on good thatched roofs using a special reed grown for that purpose. You know how "fuzzy" basswood is? Well, the inside of the bark, soaked for half a day in hot water, then pounded, makes a pretty good paint brush. The Indians stripped out that inner bark, twisted it, and wove it into cloth.

I once read that Indians used the inner bark of oak trees as diaper material. Supposed to be especially good because the tannic acid in the oak was good for diaper rash. (A tea boiled from oak bark is good for blisters.) Well, anyway I had to experiment with that bit of knowledge. I had some oak bark, a bunch of it from a dead oak. I shredded it with the lawn-

The crosscut saw is a good way to cut up trees, if you've got energy to burn. A chain saw is a quicker, less strenuous, more noisy alternative.

mower, and sure enough out came a reddish material wondrously soft and absorbent. As good as any "northern tissue" on the market and softer than any downy diaper I have handled!

If you intend to cut wood for your fuel and thereby save yourself lots of money, you will naturally cut next winter's supply this winter. December and January are the best months to do it. I made wages one winter cutting fireplace wood. Most of your wood for urban fireplaces you will sell during the Thanksgiving-to-Christmas season. You haul during that time; in January and February you cut for the next year. It is very hard work—but I have spent worse winters.

Food from the wild in January is not easy to find: best bets are what you get from hunting and fishing. Ice fishing, at least where I did it in Minnesota for several years, always brought us all the fish we cared to eat during the winter.

If you happen to live where the wild watercress grows (we do!), you can sometimes feast on one of the best salads anywhere—right in the dead of winter. If you have a good spring with a steady stream flowing from it, plant the stream to watercress.

February: By the end of the month, the sap is running in the maple trees, and whether you have two trees or two hundred, you should make a little syrup. There is nothing like it, poured over your own freshly ground cornmeal fritters or freshly ground wheat pancakes. Abundant information is now available on getting sap out of maple trees. Sugar maple is the best variety to tap—that's the maple that sports clusters of tiny yellowish-green flowers in early spring. I would mention further—since tree-lovers bring it up—that university

Tap a maple, if you have one. It takes but a bit of effort in a slow month, February, and the reward is tasty maple syrup for your pancakes and maple sugar for munching.

experiments say no harm comes to the tree from drawing sap out of it in early spring.

A hole of about one-quarter inch in diameter is sufficient to draw sap from the tree. I make mine large enough to take the spiles I make out of sumac wood—pieces of branch about a half inch thick and six inches long. You can push the inner pulp out of sumac rather easily to give a hollow tube. Of course modern metal spiles are much better; I just happen to like to do a few things for fun the old way.

I don't use my wood to boil off the sap though. Takes too much, and too much I rarely have. Instead I'll use a kerosene or oil burner with a broad shallow pan. If you are only going to boil down a couple of gallons of sap for a "taste," you can do it on the stove in the kitchen. Anymore than that, however, do outside. All that steam in the house can play heck with stuff like wallpaper.

March: In the south, "poke," the poor man's asparagus, breaks through the ground and by the end of the month you can find it as far north as the Ohio River. If you don't know what poke is, forget it, or find someone who does. Don't try to rely on books to identify wild plants.

There are other spring plants that are good to eat, but the dandelion is the favorite of most "yarb-hunters," probably because it is easily recognized and grows almost everywhere. Wilt your dandelion greens and chop in bacon and boiled eggs with vinegar. Very good before warm weather gives the leaves a somewhat bitter taste.

April-May: While you're eating dandelion greens and waiting for the asparagus to come up, enjoy early spring by rambling through the woods and fields. Your cultivated land, as yet unplowed but washed by winter's rains, is perfect now for hunting Indian relics. In the fence row, look for wild asparagus. Along the creek, mark the lay of that big rock, moss-covered on one side, that will look good as a lawn decoration. (Suburban folks sometimes pay plenty to get one in their yards.) In the woods, strip off a few bags of bark from the shagbark hickories to burn with charcoal at summer barbecues. The hickory smoke adds a special flavor to whatever you're cooking. Along the river, swollen with spring rains, watch for pieces of driftwood that you can use in your home or sell to specialty stores.

When mayapples open their "umbrellas" and until dogwoods drop their blossoms, nature offers the sharp of eye one of her most delectable treats—the morel mushroom. There are several types of morel, but typically, it looks like a piece of sponge in the shape of a miniature Christmas tree—from an inch up to six inches tall. If you are unfamiliar with the mushroom get a seasoned hunter to show you some.

Morels grow under pin oak trees, elms—especially red elms recently dead, white ash, some old apple and pear trees, and cedar trees. Actually, I've found them under almost every kind of tree on occasion and in cutover timber of all kinds. Morels sometimes grow along railroad track embankments, for what reason I don't know.

Early morels are small and greyish; later ones tan and yellowish. There's a very early small one which we call the leaf-morel in Ohio. Finding the leaf-morel is for experienced hunters only. There are other kinds with long stems, but I advise leaving them for the veteran, too, as they look too much like toadstools for the amatuer to distinguish.

Morel hunting is very popular now—too popular I keep telling myself, so why don't I quit writing about it? Soon *everybody* will be in the woods in spring; am I being selfish to cringe at that thought? Iowa recently declared it permissible for visitors to pick morels in its state parks. Michigan hosts hunters from all over the nation in May. In fact, several Michigan areas advertise themselves as the morel capitals of the world.

You can eat morels fresh—fried in butter with bread crumbs is the best way, though some people use a cornmeal batter—or you can dry and then freeze them. Reconstitute them with a little water when you thaw them for meal preparation.

June: Wild strawberry time and no berry tasts better. Patches of wild strawberries do not produce consistently; last year's mother lode may yield nothing this year. But if you have marked the location of several patches during year-round hikes in fields and along country roads, you have enough alternate sources for an occasional dessert or a few jars of jam.

Before highway departments started spraying roadsides with weed killers, wild strawberries grew in abundance along the backroads of the east and midwest. In some areas, roadsides are still the best place to look. Railroad embankments are also excellent. And in the east, abandoned pastures sometimes become literally covered with the small red gems.

The trick is to pass up patches where berries are only pea-sized in favor of patches where berries are marble-sized. In this way, you double the quantity of berries you can pick in a given time—important because harvesting wild berries can be slow going. If you wait until midseason, most of the berries in a given cluster will be ripe. Then pick the whole cluster by the stem even though a berry here or there is unripe or over-ripe. This harvesting method speeds picking time, too. And if you place whole clusters of berries in your picking box, you won't mash the berries as you will if you fill a box with the individual, fragile berries.

July: Now you can add to your larder of fruit by harvesting wild black raspberries. In Michigan and Ohio I have also found wild yellow raspberries, so keep an eye peeled for them. If you see a bush where all the berries seem to be yellowish-orange with no trace of the red of immature black-caps, you've found a yellow raspberry. When ripe, it will, like its brother blackcap, slip easily off its center core.

In a good year, wild raspberries produce berries almost as large as domestic black raspberries. Since the fruit is common in all the states I have hunted it, I usually cruise right past patches where the blackcaps are small and seedy.

We do more woods-walking in early spring and late fall than in summer. But you can easily spot and mark for future reference, patches of black raspberries in their offseason. The thorny vines have a distinct purplish hue to them.

August. Round out your supply of wild fruit, jams and jellies with blackberries, usually more plentiful and faster to pick than raspberries and strawberries. (Wineberries, which look like orange-red raspberries, and wild blueberries and huckleberries are ripe now, too, if you are lucky enough to have them in your area.) Blackberries are, without doubt, the most practical wild berry to harvest because they are the most plentiful. Time was, my mother-in-law would take her daughters to the woods in August to pick blackberries by the gallon to sell on the city market for money to buy school clothes. But that was a long time ago, and though the consumer would still pay well for good quality blackberries, most kids won't work that hard in August heat. And those who used to do it do not recall the memory fondly.

But for your own use, the job is not at all formidable. Here again, be choosy. Scout around until you are satisfied that the berries you are picking are the biggest, juiciest ones available. Some small to medium-sized blackberries have a hard core, while bigger types seem to have little or no core at all. The smaller ones aren't much good for eating fresh, but they make good jam or jelly.

Forget not blackberry pie, blackberry cobbler, and blackberry wine, if you're so inclined. Blackberries are not only good, they are healthful, especially in late summer when you are stowing away mildly laxative foods like peaches, melons, and sweet corn. Blackberries contain an ingredient to counteract what we used to call summer complaint. In fact in mountain folklore, a tea made from blackberry roots is considered the cure for dysentery.

Elderberries ripen a little later than blackberries. They grow on high bushes in the same general areas that blackberries do. It is easy to gather the black clusters of small berries by the stemful and fill a basket in a hurry. Back home, you pull the BB-sized berries from the stems and make elderberry pie or elderberry jelly. The jelly is the best there is to my taste. The berry makes one of the traditional home wines, too. I guess I can't make it correctly; mine always tastes a little like kerosene.

I suppose elderberry time has special significance to me. On the old-fashioned farm of my boyhood, August was the time (as the rhythms of nature dictate) for easing up on the work, slowing down, sitting in the shade. The modern farm with its time saving devices seems to have hard work right through August. But back then, we slacked off in the middle of August, and my father would decide to go hunting for turtles or crayfish or groundhogs or carp—all of which would end up in memorable meals.

The turtles made some of the best dishes I've ever tasted (not soup; that stuff is for tourists), but I don't set lines for turtles nor tell anyone else how to do it. Turtles are slow growers, and if more people found out how delicious they are, snappers would become extinct in five years, I'm afraid.

For crayfish (which we called crabs in Ohio), we took the minnow seine down to our creek and scooped them out by the

Snapping turtles make good meals, but they are better left uneaten. They grow slowly and could easily become extinct if too many folks dine on them.

bucket full. You can't do that any more on that creek because it dries up every summer now, destroying the many kinds of fish and other water creatures that used to live there. The "Soil Conservation" Service wrought the change when it ditched, tiled, and surface-drained every tillable acre it could, so that excess water would drain off fast. It sure does. It drains off so fast that the Soil Conservation Service now feels it must build dams in river bottoms to stop flooding. Now we not only have dried creek beds all summer, but lakes inundating thousands of acres of prime bottom land farms, all in the interest of what government has the effrontery to call "soil and water conservation."

Nonetheless, if you know where you can still find crayfish, their tails taste just like lobster tails and you fix them the same way. Young groundhogs, fixed like squirrel, will keep you alive. Carp, at which some fisherman turn their noses up, is delicious if smoked slowly for 24 hours after soaking overnight in salt water.

September-October: As the year slides into fall, it's time to go nutting. There is some practicality in gathering your own nuts, but the chief advantage is that the project is such a pleasant way to spend an autumn day. Black walnuts and hickory nuts are the two varieties most commonly hunted and gathered in the north, with hazelnuts, butternuts, and beechnuts also possibilities.

Wild hazelnuts grow on bushes and are very difficult to spot unless you know what you're looking for. I've never found them in sufficient quantity for practical gathering, but they are most delicious. Butternuts are related to walnuts, but are oblong rather than round. Their husks are slightly sticky. Beechnuts are too small to gather in quantity, but extremely tasty. They are difficult to crack, too. We used to roast them in the fireplace of our cabin, which was surrounded by beech trees. As good as roasted chestnuts.

Speaking of chestnuts, the tasty American chestnut is not quite extinct. Some trees, almost dead from the blight disease, continue to send up shoots which develop into small trees. These trees sometimes fruit before the blight kills them. On the Appalachian Trail in eastern Pennsylvania, we harvested about twenty chestnuts from such a tree. Subsequent investigation brought to light other trees that were bearing a few nuts. Before I knew that, I would have liked someone to tell me—so now I tell you.

Walnuts are the most practical nuts you can gather. Generally speaking, the best way to gather them is when their husks are green. Take a sack of them home, spread them on your driveway, and let the cars run over them awhile. The tires smash the husks away without cracking the nuts, and the sun dries the latter. In about a week, you can pick up the nuts minus the husks without staining your fingers too much. Then spread the nuts out again in a sunny, dry spot (a shed roof is perfect) for another week or so. Watch out for squirrels though; they will rob you blind, given half a chance. Then you can store the nuts for winter use.

In gathering hickory nuts, the taste of which comes straight out of a paradise that kings and queens can only pine for, it's best to wait after the first good frost, when the nuts will have all fallen to the ground or can easily be shaken down, and when the outer husk falls freely from the nut. However, if squirrels are taking all the nuts from your favorite tree, you may want to harvest sooner. Let squirrels eat acorns, I say.

Always check the quality of the hickory nut from the particular tree you intend to harvest. If the nuts are mostly wormy, go to another tree. If the nut is thick-shelled with a small amount of meat, try another tree. Assuming you have some choice, and you should have in most parts of the north and east, you will usually be able to find a good shagbark hickory with thin-shelled, large-meated nuts. That's the tree to spend your time with.

If you spot a tree that looks like a hickory, but with nuts that have thin husks that cling to the nuts even after they fall to the ground, you are probably looking at what we called "pig nuts." They are more bitter and of a lower quality than hickory nuts.

You may find some hickories with nuts the size of a ping-pong ball. These are "bull-nuts"; the shell is usually thick, but the meats as good as any hickory.

Cracking the meat out of the nuts is time consuming; but much faster with walnuts than hickory nuts. You can more quickly crack out a cupful of nut meats if you crack a panful of nuts all at once, and then pick out the meats, rather than cracking a nut, picking out the meat, and then going on to the next one. We make nut cracking, like nut gathering, a family affair. You might be surprised how much communication between generations can develop when the whole family sits around the table picking out nut meats from the shells. And once the younger generation tastes a hickory nut pie or a black walnut cake, I'll bet you won't have much trouble getting them to help either. Incidentally, while I have no special way to crack a walnut, a hickory nut will split to allow you to take out two "whole halves" if you stand it on edge and strike the top with a hammer. Most of the time.

During the nutting season, keep an eye out for wild grapes too. They look just like Concords only the grapes themselves are smaller in size. They make excellent jelly and good wine, though the latter is too "foxy" in taste for some wine fanciers.

Fall is the time to gather bittersweet and all sorts of dry or driable plants for decoration and bouquet arrangements. We are able to sell sprays of bittersweet very easily and used to have a nice little "pocket money" trade going. Many kinds of grasses, weeds, and cattails make marvelous vase arrangements.

November: Many veteran countrymen dread the coming of hunting season in this month, when so many irresponsible gunners carry their weapons and their whiskey into field and woods to spend the days blasting everything that moves including each other. But controlled harvesting of certain species of wildlife seems to be not only appropriate but necessary. Man has already upset nature's balance in this regard. Many animals, without their natural predators and without their natural food range, overpopulate and die without controlled hunting.

Wild game is a good source of meat protein, and those who hunt in need of food have never been the ones for whom hunting laws and regulations had to be made. The good hunters take only what they need, and like the good farmers with their land, make sure there are plenty of wild animals left to provide food again the next year.

Many country families still rely on wild animals to supplement their diets. I am thinking of a wonderful but poor dairy farm family (poor in earthly possessions, certainly not in spirit) I knew in Minnesota a few years ago. They owned sixty acres and twenty-five cows from which they derived just enough extra cash to stay out of debt. The last time I was there, the oldest boy (of four children), had taken forty-five squirrels that year and about as many rabbits with his rifle. He was proud of his hunting and with reason. The family needed the meat and the dollars it saved meant, as he soberly put it, "new shoes for everyone." His father generally shot a deer each year, plus duck, geese, and pheasants. His mother worked very hard, turning the wild food, along with what came from farm and garden, into meals. But her grocery bill was almost zero some weeks. And never a whole lot. And she and her husband owed money to no one. And they were always laughing.

December: If pines or other kinds of evergreen trees grow wild in your woodlands, take advantage of them at Christmas. We always cut our own tree—a wild cedar, a variety most people wouldn't want for a Christmas tree, I guess, but we like them. You can also make garlands and wreaths for yourself or as presents for friends or to sell. Don't forget to gather a few pine cones for the wreaths. Pine cones can be sold, too.

If you have the land but not the evergreens, plant some. In a few years, you can have Christmas trees to sell. The fad of planting evergreens in the hope of becoming rich at Christmas time has died down; there's more work to it than people realized. However, you can still do splendidly selling an acre or two of Christmas trees to those who want a freshly cut tree —and not have to work terribly hard either.

December is the time to sell that firewood you cut last January—and to start cutting the wood you'll want next winter. Some of the old wood-wisdom might come in handy. Birch, fir, poplar, and elm burn poorly or too fast. Apple wood, when burned, gives off a kind of perfumy scent and old-timers always liked to throw a stick of it in the fireplace once and awhile rather than use it all up and once. Beech wood, above all woods, should be burned only after laying in the log a year. Oak, maple and white ash are about the best for fuel. My grandfather used to have a little rhyme which for all I know, may be ancient folklore:

Oak and maple, if dry and old
Keep away the winter cold.
But ash wood wet and ash wood dry,
A king shall warm his slippers by.

The drawback to gathering wild foods is that unless you are lucky enough to have a large acreage of your own, you must depend on access to public lands or permission of a considerate neighbor. In public parks, you will be subjected to all kinds of rules—you hardly dare pick a dandelion in parks anymore. In the second case, private landowners may allow you to gather wild food on their places or they may not. And in either case, other hunters may get there before you do.

This is one of the reasons I advise the prospective organic homesteader to try hard to either buy a place that has a pond and a woodlot on it—preferably adjacent to each other—or buy a place where a pond and woodlot can be added. You can actually accomplish this feat on fairly small acreage. A quarter-acre pond and an acre woodlot will suffice to bring all the wildlife I've mentioned to your doorstep.

A farm pond is the single most enjoyable and practical addition you can make to your homestead. It becomes a center for the observation and study of literally hundreds of kinds of birds, bugs and animals lured by the water. It becomes a haven for your body and soul. It becomes, in a very real way, one of the few sources of water you know is unpolluted.

You can have fish, turtles, frogs, snakes, and crayfish in the water; ducks and geese on the water; many varieties of songbirds above the water; mink, muskrat, raccoon, oppossum on

A farm pond can be a productive and enjoyable part of most homesteads, since one can oftimes be constructed where none exists.

the banks; and maybe deer for a drink of water. There's boating, swimming, and fishing in summer; trapping, ice fishing, and ice skating in winter; and picnics the whole year.

The "work" of managing your pond properly is half the fun. You fish out extra fish to make room for the small fry to grow bigger; you harvest bullfrogs for frog legs, and crayfish for their miniature "lobster tails." I don't advise crayfish in a pond, but if they're already in it, turtles will help keep their numbers down. If the pond gets overcrowded with edible types of turtles, you can keep their numbers down and give yourself delicious meals at the same time. If too many cattails grow up, you can dig them up and make a tasty vegetable from the roots—or let the muskrats clean up the cattails, and you clean up the extra muskrats by trapping them.

Build a pond if you can. It may cost you the price of a new car (but not necessarily—my father built one with a tractor and manure scoop for about $200), but unlike a car, it adds value to your property and does not depreciate. If handy to house or barn, it can lower your fire insurance by providing a source of water in case fire does strike. Moreover, you will be striking a blow against pollutant. Erosion, as I've said, is still the number one pollution in this country, and conservationists are beginning to understand that better water quality, fewer floods, and less erosion can result only from stopping water before it runs off. Your farm pond helps to do just that.

If you are buying farmland for your homestead, don't take a real estate agent's word for what is a good pond site. Land salesmen will advertise hog wallows as good pond sites if they think they can get away with it. Contact the nearest Soil Conservation Service office and ask them about good pond sites and whether the soil types on the land you are looking at will hold water. Once established on your land, you can get the Service's local conservationist to design your pond if you want him to, give you estimates of how much dirt needs to be moved for the dam structure, and give you an idea of the costs involved. All you have to do is become a cooperator with the Soil Conservation Service.

You can, however, be much less elaborate about the whole thing. My brother has a small pond practically in his back yard which is built up on fairly level land and filled by pumping instead of run-off water.

But for an acre or two-acre pond (and there's not much use in building a bigger one unless you just want to be extravagant) you want a ravine, a *small* ravine where the two hillsides forming it are close together so that the dam won't be so large as to be too expensive. From my experience, I wouldn't want more than twenty-five acres of land draining into a one-acre pond. And most important, the land that does drain into the pond should be in woodland or permanent pasture, not cultivated crops. If water flows over tilled land it carries the soil on into the pond and in a number of years can literally fill the pond with silt.

A spring-fed pond is, of course, best of all, if you happen to have good springs. Another idea is to scoop out a pond beside a creek and use the stream water to fill the pond. I just visited a farm where such a pond was being built. After he had bulldozed out the pond, the farmer dug a ditch from pond bank to creek bank, into which he put an eight inch water pipe with a valve at the upper end. Then he dammed up the creek, the water rose to the pipe inlet, and flowed on into the pond.

Creek-side ponds may have a weakness that could stop you —often the ground in such an area is underlaid with gravel and the pond won't hold water very well.

For a regular pond fed by run-off water, you will need engineering help not only on dam construction but to determine the size of your overflow pipe and emergency spillway. You don't want heavy rains to cause your pond to rise over the dam and wash it away.

Once you have a pond, the first thing you want to do is stock it with fish. As far as I know, wildlife technicians are still advising a mixture of bass and bluegills for warm water ponds. I would strongly urge you to stock *only* bass. The bluegills will overpopulate in almost every case, despite what the wildlife people say. But even if you were lucky, and the bluegills wouldn't overpopulate, why take the chance? Bass are more fun to catch and tastier to eat. Stock your pond with largemouth bass only. And after a couple of years, fish them hard. Don't throw small ones back in; eat them. You see, fish easily overpopulate, which gives you a pond full of small fish which never grow large enough to give you exciting fishing. The bass–bluegill mixture is supposed to give the bass something to eat—bluegills—and thereby keep the latter in check. But bass will eat their own young, which is better, because that keeps population of bass down where you want it. Pole fishing alone seldom "prunes back" the fish population adequately, especially if you stock both bass and bluegills.

If you have a cold water pond—one where water temperatures stay in the fifty to sixty-five degree range and never get higher than seventy-five degrees—you can raise trout. Rainbow and brook trout are the two best kinds to stock. Don't stock other fish with trout. Your fish and game agency or the local SCS office can give you a list of hatcheries that sell fish to individuals.

Most artificial ponds will raise largemouth bass much more safely and surely than trout—water temperature being the determining factor. Artificial ponds in the West above five thousand feet do, however, usually stay cold enough for trout.

If you want some other variety of fish with largemouth bass in your pond, stock channel catfish rather than bluegills. The channel cats may not reproduce themselves, which may sound like a disadvantage. But better to restock every few years than to get too many fish. Regular stocking rate for channel cats is about one-hundred to two-hundred fingerlings per surface acre. If big bass are already in the pond though, stock fewer channel cats of a larger size—five inches to eight inches. Bass will eat smaller ones.

Don't let anyone fish your pond with minnows for bait. That's a good way to get unwanted species—like carp.

Raising catfish in ponds has become an important commercial enterprise, especially in the South. It is not, however, the easy-money project some bulletins make it out to be. If you are tempted to try this business, be sure to visit the catfish farmers in Alabama or Mississippi before you invest a lot of money.

The main problem with artificial ponds is weed growth. Shallow shore lines and shallow upper ends of ponds will grow water weeds that ruin fishing and make your pond unsightly. Shore lines should be steep. Pond inlets should be deepened at the time of construction.

Algae and other floating weeds are a problem, too. They not only cause unsightliness and stop fishing and swimming, but they can rob enough oxygen from the water on cloudy days to kill fish, especially trout.

There are chemicals to control water weeds. Lots of people have killed more than weeds with these chemicals—like, for instance, fish. But organicists will hardly want to use chemicals anyway. Instead, on a smaller pond, you can quite practically clean out floating weeds by gathering great gobs of them into your arms as you wade along. The reason this chore is worth your while is that the weeds make excellent mulch for your garden.

A colony of muskrats will clear a lot of water weeds out for you, too. You have to watch muskrats closely though. If your dam is not quite as high or as wide as it should be, muskrats sometimes tunnel into it and come out on the other side. The water follows the muskrat, and in less time than you think, the whole center of the dam can wash away. I've seen it happen.

Your pond will lure mallards and Canadian geese if you are located along a fly-way. Even if you need extra meat, I don't advise shooting Canadian geese. They just don't taste all that good. Besides if you are nice to them and don't live too far north, you may very well entice a pair to stay on the whole year round. The geese will become almost, but not quite, domesticated. We have a whole flock of them on a private pond just a mile from where I'm writing. But no gun is ever allowed there.

If you put up special nests for wood-ducks, you may get a pair of them—the most beautiful of all birds to me. But don't count on it. Wood-ducks are shy and often won't use the nests which wildlife experts build for them.

Buy a pair of bullfrogs or catch some from another pond to get these insect feeders started on your pond. I don't know exactly how many bullfrogs can live, or will live, in a half-acre pond. But you might as well harvest some of them every year

The bullfrog is to the pond what the toad is to the garden, a real friend. He'll eat most anything within reach that moves: insects, spiders, minnows, crayfish, even smaller frogs. As such, it's worth the trouble to establish a pair at your pond.

after they get established, because the frogs won't tolerate crowded conditions and after awhile, the young ones will leave in search of more elbow room.

Your pond has another practical use—as a source of supplemental irrigation water. Often though, the time you need to irrigate is also the time your pond is at its lowest level. In other words you can't always depend on irrigation water from a *small* pond to be there when you need it.

The next best thing to a pond is an unpolluted stream, which if large enough, offers many of the same food benefits and pleasures that a pond does. You can't very well stock the creek with fish and know they will always be there—high waters will carry them on downstream from your property sometimes. But if the stream is large enough and clean, it will usually have fish in it naturally. If you stock a few more, you'll get some benefit from the effort.

If you have a small homestead—too small for a pond and not bordering a clean stream—you might be interested in raising fish in small, man-made tanks and pools. This activity is receiving considerable attention at present. There seems to be no good reason why a person can't raise fish as easily, or even more easily, than rabbits or chickens. Makes sense to me anyway because when I was a boy on the farm, we always kept fish in our four foot by four foot by eight foot cement horse trough. Catfish seemed to adapt to that environment very well. They grew to edible size and even spawned a couple batches of young.

At least one commercial company is advertising a system it claims makes trout raising as easy as growing vegetables. (Life Support Systems Inc., 5405 Gibson Boulevard Southeast, Albuquerque, New Mexico 87108. The president of the company, Gene Bussey, has written a book, *How to Grow Trout in Your Back Yard.* I have not read the book nor used the system so cannot recommend either. I pass it along as interesting information. Bussey claims that using his method, you can raise a thousand pounds of fish at a time in a four foot by four foot by eight foot pool.)

The most exciting experiments using fish on a small organic homestead are being carried on by the New Alchemy Institute. You may have read articles on the subject in *Organic Gardening and Farming* magazine, authored by John Todd, who works at Woods Hole Oceanographic Institute in Massachusetts. So far, the main objective of the experiments has been to find the best kind of fish to use. Seems to be a semi-tropical variety of Tilapia. But testing goes on, and all homesteaders are invited to help. The idea behind the experiments is not simply to raise fish, but to combine fish raising with a homestead that recycles all wastes. In some parts of Asia, fish farmers use ducks and field crops in a sort of total polyculture of which fish are only a part. The ducks, raised on the water, fertilize the pond with their droppings. When nutrients in the pond get too rich, the water is drained out, and the soil spread back on the fields. Pond water itself, if occupied by a lot of fish, seems to have much more fertilizing power than well water or rain water.

My next move on my homestead is going to be a halting and much less romantic step toward fish-raising. I've thought about it quite awhile, which means I've been trying to figure out a way to raise fish without spending much money. Remembering the horse trough back home, I know I could build one like that (cement) for perhaps $50 and a lot of sweat and muscle power. Before I expend that kind of time, I am going to risk $50 with an easier way: a metal cattle trough from an Agway farm supply store or from Sears. The troughs come in varying sizes, but the one that sells for $52 at Agway is about three feet by three feet by eight feet. I am going to set the trough up by the barn. One of those pumps that you can buy to make a fake waterfall would suffice to keep the water circulated and oxygenated, but I want my tank to work without electrical appliances if possible. So I am going to make one of those wind-powered water stirrers cattlemen use on ponds to keep the water from freezing. Sears has these stirrers in their farm catalog, if you are interested. I will rig up some kind of paddle on the other end of the wind-blown blades, which will splash the water around enough to stir in some oxygen. Oxygen is the limiting factor—well, the first limiting factor—in how many fish you can grow in a tank. Into my tank, I will also funnel all the water that comes off the barn roof in rain. That will supply a periodic shot of fresh water to the tank. I'll start out with two adult catfish and four adult bass, and see what happens. Wish me luck?

Chapter Ten

Making Money on Your Homestead

For the kind of organic homestead discussed in this book, I see only three ways in which I honestly believe you might make money. All three derive indirectly from your main purpose in homesteading, which is to gain a greater independence and self reliance by providing as many of the necessities of life for your family as you can in a manner which brings you a greater quality of life. That's more important than making money.

—1. You "make" money by living organically because you live life sanely and therefore pleasurably. You *enjoy* what you are doing. How much is that worth? Because you are living at an easier, but happier pace, you'll live longer; your work keeps your body in trim and your mind busy but unharried. The food you raise is better than what you can buy and better for you. What's a long and more satisfying life worth? Moreover, when you are hoeing a garden you cannot be spending money on more faddish and fashionable pursuits that demand outlays of cash. Nor are you, while hoeing, risking your neck on the highway. How much does hoeing save in dollars?

—2. The second way you profit monetarily from your

homestead is in the seemingly irreversible rise in land prices. You usually have invested in more land than the ordinary suburbanite, and in almost every case, your investment will be the best possible hedge against inflation.

——3. If you manage the production from your garden with old Yankee wit and parsimony, you will save an appreciable amount of money. The reason is not really that you do all that work yourself (in strict financial accounting, you should charge your labor against your gross profits), but that you can legitimately "charge" yourself the going retail price for what you consume from your garden and barn. You would otherwise have bought that food in the store, so that's what it's worth to you in cash. In others words, you "sell" what you eat at retail price without bearing the cost of retail merchandising. Any commercial farmer who could do that would become rich very quickly.

Beyond those three considerations, I do not want to leave the impression that you can make a good living or even a major part of a good living from a small organic homestead. You *can* do it, of course—a couple of acres of greenhouses, a craft shop, a small factory, a bakery—but if you undertake to go into business full time, then your homestead really becomes an adjunct to the business, and the business side of it demands much sterner dictums of methods and financing and know-how than I have set forth here. Commercial farming is no way, no way at all, to escape the rat race.

But you can make a *little* money, and your youngsters growing into their teens can make a little money—and learn the pride of accomplishment that comes with contributing to a family income. And while you are making it, you will learn something about commercial farming and be in a much better position to decide whether you wish to move gradually in that direction.

By any standard, as an organic homesteader, you will be a

very small producer of food or other goods for marketing. The only way you can pocket any appreciable profit at all is by selling retail—direct to the ultimate consumer. On a spare-time, part-time enterprise you can accomplish this by a temporary roadside stand, or by delivery to a regular list of customers, or simply by selling to a select group of friends who stop by your place when produce is available.

Your temporary roadside stand need be nothing more than a card table with an umbrella over it to shade out the sun. You can sell on weekends only, or whenever you feel like it. You will get some business—I have never known a roadside stand not to get at least a little business no matter how lackadaisically it was run. A weekend roadside stand can be enjoyable simply because you're doing it for fun and not worrying that you're going to go broke. Teenage sons and daughters often find they enjoy running the stand—at least for awhile. Some very successful roadside stands started when fathers suggested to daughters that they sell extra garden produce to passing motorists. In one of these cases, the roadside stand business finally eclipsed the father's dairy. He quit milking cows and threw in with the girls!

Whether you keep your stand open all the time or only when you feel like it, there is one rule you must follow: sell only fresh, quality produce. If you don't you will lose the few customers who do stop by, and they will tell others their bad experience. If you have true organic food, make a point of it. Most knowledgeable consumers prefer buying fruits and vegetables that have not been sprayed with poisons. Finally, don't sell your food cheap. Charge what the supermarkets charge. If your stuff is fresh, people will be glad to pay the proper price. You are proud of what you have raised; customers know that when you put a decent price on it

If you want to give away your extra produce, you can make even that pay. Your friends and acquaintances will appreciate a gift from your garden or farm much more than a store-bought trinket at Christmas, or on anniversaries and birthdays. A basket of fresh vegetables for Uncle Joe; a big spray of red-berried holly or a homemade wreath for brother Ed and his family at Christmas; a jar of homemade wild strawberry jam or some genuine maple syrup you made yourself for sister Jean. They'll be thrilled—and you will have saved no small fistful of cash.

Another way to make a little money on the side is to aim your sale at specific holiday markets: sell iris and other flowers before Memorial Day; daffodils, chicks, baby rabbits, eggs at Easter; pumpkins for Halloween; Indian corn, gourds, bittersweet, and more pumpkins for Thansgiving; holly, Christmas trees, evergreen garlands, and wreaths at Christmas. Flowers sell well for almost all holiday and anniversary seasons.

Specialty selling may be your way to dip your toe into commerical production. If you enjoy growing a particular kind of fruit or vegetable and find by experience you can do a good job of it, you can concentrate on growing that crop for sale. One gardener I know makes a specialty of beefsteak tomatoes; another, a certain large-sized muskmelon; another, old shoe-peg Country Gentlemen corn. Each has established a reputation in his neighborhood for his specialty, and he has standing orders for all he cares to sell. Some farmers smoke hams and sell them the same way. All of these projects take up a minimum of time—rather just that amount of time the fellow wants to put into it. You'll have no trouble disposing of your specialty once people find out how good it is.

Some gardeners, in their retirement years, make arrangements to supply nearby restaurants with high quality vegetables or fruits. Restaurant owners aren't as likely to buy from you as they once were, because they must have a steady, year-round supply—a service which a commercial food distributor can guarantee and the gardener cannot. So the restaurant does business with the distributor even if his vegetables aren't as good as yours. But some restaurants may be tickled to get good quality food in season. You have to talk to the owners and bring along proof of your claims for the chef to sample.

My own method of making a little spending money from our homestead is built around five egg customers. They appreciate fresh eggs enough to drop by for a weekly dozen, so we know they appreciate fresh fruits and vegetables. If and when we have extra, we let them know it's for sale without pressuring them. Actually we give some away, too—it pays to cultivate discerning customers.

When I retire and expand my "commercial enterprise," I will do it by adding more hens to the flock, more customers to the egg list, and gradually show them how I can supply much of their produce. My service will be limited to a few families, and I will not have to be "open for business" to the general public, the way you are with a roadside stand. I can't think of an easier way to make a little extra money.

Another kind of specialization which has some advantages is in selling some food you have *processed* from your crops. Homemade jams and jellies come to mind immediately, because I know of homesteaders who have successfully marketed them. First of all, people like homemade jams. Secondly, you can make them more or less at your convenience, and then can sell them at your leisure. Thirdly, and this applies particularly to organic growers, you can use in jams and jellies blemished fruits which you can't easily sell fresh because they don't have that false perfection of sprayed fruit. You can cut out worm holes and blight scabs and use the unharmed parts in your jams. Fruits that get a little too old for fresh sale still make dandy jellies. And lastly, at home you can make many kinds of preserves that are hard to find on the commercial market: elderberry jelly, wild wineberry jam—there's even a recipe for corn cob jelly!

Vegetables, too, can be preserved for sale as corn relish, chili sauce, horseradish. One farmer in Chester County, Pennsylvania, has made a good living growing raspberries and processing his own horseradish for sale in stores.

Honey from your own beehives is another sideline from which you can make a little money. Selling fishworms to fishermen requires only a small amount of your time and can be profitable too. A small greenhouse can be the start of a spare time, profit-making hobby, especially if you use it to raise rarer forms of houseplants. All you have to do is read classified ads, and you will find half a hundred ways to turn spare time activities into a little profit on your homestead. But don't allow yourself to make the mistake of thinking any of them will make an appreciable amount of money. Some might, but only if you turn them into full-time business occupations.

If your homestead is a larger one with a nice forest and/or pond on it, you can "rent" it to urban people at specific times for fishing, picnicking, or perhaps even swimming. This practice has become a regular business with some farmers, some of whom have gone one step farther and developed camping grounds for travelers. But I don't advise you to open up to the public unless you have found that such a commercial enterprise appeals to you. Instead, keep your eyes and ears open for the occasional city family who would gladly pay a fee for a *private* picnic and/or fishing spot they could use occasionally. Be sure you are properly insured. Have a list of do's and don'ts these "renters" should follow as part of the deal.

I should stop discussing ways of making money through farming right here, but I know the reader with true homesteading blood in his veins would be disappointed. So I offer a few examples how a person *with experience* might work into a small but profitable farm operation without investing the huge sums of money typical commercial farming takes today. But I must stress that a small farm providing a good living for a family in a sort of old-fashioned, unhurried, un-rat-race kind of way is a utopia more often written about than actually achieved.

From a lifetime of living on farms, working on farms (in six different states at that) and plotting how my own version of that fleeting, aforementioned utopia, I have selected five basic requirements a small farm operation would have to meet to make the grade and fit my personality at the same time.

——1. The farm must provide a net income of at least $10,000 a year. I wouldn't settle for less in this day and age and would hope for more.

——2. The operation must not involve big capital investments. Not only couldn't I afford them, but even if I qualified to borrow a lot of money, my temperament is such that I just don't *like* to owe big money to banks and pay big interest for the privilege of owing it. If I did, I would have been a businessman long ago.

——3. The project would not involve daily contact with the general public. I like people, but I could not endure facing customers every day in a retailing business for very long.

——4. I would have to be the boss, but I would not want to boss other people. In other words, the farm would have to be able to be managed by my family and me.

——5. The particular enterprise of the farm would have to be one I like to do and know how to do.

Perhaps not all those requirements figure in your idea of utopia, but I have a feeling they do apply to most people interested in an organic homestead. At any rate, finding a farm enterprise that fits all five is, for all practical purposes, impossible.

Those farm operations that come close—that net (before

living costs) $10,000 or more without big investments in land, machinery, or livestock—are only possible when the operator is willing to substitute time and labor for machinery and perform tasks most people find unappealing. A Maine farmer who lives much in the way this book advocates says it much better: "If you can learn to enjoy physical labor and lots of it, you can be free without being rich."

The two ways a small farmer can make a living without big investments embrace either this willingness to use physical labor in areas where machines have not been invented to take man's place (as harvesting berries) *or* in selling his farm production directly to the consumer, that is, retail.

That means that a roadside stand remains, even today, the best way for a small family farmer to make a living from farming. Often, the running of a roadside stand combines both of the requirements mentioned in the preceding paragraph. Many of the products you sell, especially berries, involve a lot of handwork that machines can't do very well yet, either in planting, pruning, harvesting or delivering (quickly) to market. A successful roadside stand operator, if he is small and starting poor, should keep that thought in mind when deciding what he should sell: what can he do better than machines? He'll grow asparagus and sweet corn because he can sell it fresh out of the field, the only way these two vegetables taste good. He'll grow good dessert-type strawberries, like Fairfax, which taste very good, but which don't ship worth a darn. He can compete with big California growers on these terms. He'll grow and sell raspberries because as yet, no one has come up with a practical machine harvester, and because raspberries don't ship well, either. Furthermore, raspberries are one fruit the northerner can grow better than the southerner. Very, very rarely can Florida or California send raspberries to northern markets ahead of the season, and wreck the market as they do with their other green-picked, tasteless produce. As a matter of fact, you *can't* pick a raspberry until it's ripe. It won't come off its stem.

The roadside stand farmer will concentrate on produce like muskmelons, big, juicy varieties of tomatoes, and peaches, which are not good if picked green for distant shipping, but which, when vine or tree ripened, can only be harvested efficiently by hand. The smart operator will not grow ordinary snap beans which can now be grown commercially in large acreages by machine. But he might work up a clientele for the longer and tastier pole beans, which require hand harvesting. Nor will he concentrate on apples or potatoes which can be handled easily now in large acreages, unless his local conditions dictate otherwise. I see little sense in growing watermelons, or trying to, in the North because they do not grow well there, and southern melons shipped in will beat you every time.

Try pears instead of apples; lots of people in the Midwest and the East don't know what a good pear tastes like. However, in selling to the public that very fact can be a drawback. If people are not used to eating pears, then, by hickory, they won't buy pears. You have to do a selling job on them. The only way to do that is give them a free one to eat.

Try eggplant instead of cucumbers. Just about anyone can grow a surplus of cucumbers. Or try some of the finer, newer kinds of squash, like Golden Nugget. The new kinds don't grow as easily as some of the older varieties, but it might put you one up on the other roadside stand operators.

The roadside stand's advantages are its weaknesses, too. You have to like the public to retail successfully. You need to know as much about retailing as any retailer—or be smart enough to learn fast. You are actually combining the talents and expertise of farmer, processor, packager, salesman, and retailer. And almost always you will have to become a labor manager.

A roadside stand is your best marketing approach if you plan to make vegetable growing a source of income. A successful stand need not be this elaborate.

As you grow in size, if you do, you will have people working both in your stand and in your fields. And labor can be a big problem today. Especially field labor. Your best bet is to hire school-age kids, but they are not very dependable and the government has a whole barrel of regulations concerning minors employed on farms. I would not even think about a roadside stand unless I either had figured out a way to do it with only family labor or knew exactly where my outside labor was coming from.

Selling retail at your own stand, you should be able to net $1,000 an acre from berries and some vegetables. So, ideally, you could net a $10,000 annual income this way with only ten good acres. I advise you to shoot for something like that (though few years will you actually make it), and instead of expanding the farm end of your business farther, buy other garden produce in your area wholesale and sell it retail. If you like this kind of business, I think that is smarter than expanding an operation to the point where it demands a lot of hired field labor. If you have a good imagination, once you start your roadside stand and have things moving along okay, you will think of a hundred different ways to expand sales without necessarily expanding hired help.

I've said it before, and I'll say it again: grow asparagus, strawberries, raspberries, blackberries, sweet corn, muskmelons, big tomatoes, squashes, and pumpkins as the crops to make your money on.

If you don't like to raise fruits and vegetables, does that mean you can't make a living from a small farm? No. You can raise flowers. You can produce anything that yields a high per acre income the way fruit does. When you get into the area of grain farming, you are talking about farms that are very large and require machines that cost $20,000 each. Today small-scale farmers just can't afford to rely on the staple grains for a livelihood.

If, however, you own or can afford to buy three hundred to five hundred acres and the necessary machinery, you will do about as well as if you invested the money in stocks and bonds. If you are trying to figure out a way to farm that much land organically, I suggest that you consider a program of double-cropping wheat and soybeans where the growing season is long enough to ripen both crops. A little farther north, double-cropping barley and soybeans might be even better.

Commercial farmers have already proven the practicability of double-cropping wheat and soybeans, though the advantages have never been emphasized for organic farmers. The wheat (or barley) is planted in the fall, October in my part of the country. The wheat, as I explained in an earlier chapter, ripens in July, and the soybeans are planted immediately after the wheat is harvested. In turn, the soybeans ripen by Octo-

You'll need this much space and more to make a go of commercial grain farming. The expenses are high, the labor requirements relatively low, the market chancy.

ber, and wheat is planted immediately after the soybeans are harvested.

In terms of labor, you have to work only two months. (July and October for the barley–soybean rotation in my area, June and October for the barley–soybean rotation. Barley ripens earlier than wheat, so you can get your soybeans in the ground quicker, which means a better yield and possibly an earlier harvest.) This advantage opens the possiblity of *renting* the machinery you need for the entire operation instead of owning it. Not only might it be cheaper to rent, but even if you had the money, you need not tie it up by purchasing machinery you use only two months—a consideration extremely important in commercial farming. The third advantage from double-cropping is that the land itself is used very efficiently—there is always a crop growing on it.

For the organic farmer (and indeed any right thinking farmer) there are other advantages. First of all, the wheat or barley clothes the fields all winter and spring and the soybeans from summer to fall, so that there is very little chance of serious soil erosion. Where laws enforce soil erosion control now, this advantage can be crucial even for those who care more about profits than conservation.

Secondly, once you have raised the soil fertility level of your fields to the point where good crops of wheat, barley, and soybeans can be produced, you can keep them at that level *organically* with a lot less expense than might otherwise by possible. The reason is twofold: wheat, barley, and soybeans do not take as many extra nutrients—and I'm thinking particularly of nitrogen—as does a crop like corn, for instance. In fact you can over-fertilize wheat or barley fairly easily with chemical fertilizers, causing the grain to lodge. Soybeans, on the other hand, do not respond much to direct applications of chemical fertilizers. Secondly, in producing a wheat or barley crop, you take from the soil only the grain, and the straw is all chopped, disked, and returned to the land as organic matter. Then you follow with soybeans, which, being a legume, add nitrogen to the soil. Again, you harvest only the beans and return all the bean straw and leaves to the land. Nor do you actually lose if, in some years, early frost hits your late-planted soybeans before they ripen. You can quickly plow the beans under for a green manure crop. With a regular program of manure, rock phosphate, lime as needed, you ought to be able to compete closely with users of inorganic fertilizers.

The wheat– or barley–soybean rotation should work very well for a number of years without weed control problems. The reason? In late June or in July when you work up the ground to plant soybeans, you kill weeds which had started in the wheat. You are past the time of year when weeds make their fast growth. Planting the soybeans solid (rather than in rows) continues to hold down weeds that would grow the rest of the summer. The beans get off to a quick start and shade out many of the weeds that would try to grow. Then in the fall, when weeds don't grow at all, you plant the wheat or barley which blankets the field and makes good root growth. In spring, the grain gets an early start and stays ahead of most of the weeds until harvest and soybean planting.

What will be your returns? Wheat varies in yields around the country, depending on variety, weather, and skill of the farmer. Average yields nationally are around thirty bushels per acre, but on good land it is quite possible to double that. For the states I am familiar with in the eastern cornbelt, if I couldn't raise forty bushel wheat, I'd quit. So if you figure on a one-hundred acre basis, that's four thousand bushels. Don't figure more than $1.80 per bushel over the long haul. That would give you a gross of of $7,200. With good management you might net $3,000 over costs.

On the second crop, soybeans, you will not get as good a yield because you have to plant them later than normal, but farmers generally get from twenty-five to thirty-five bushels to the acre on such a late crop. Figure twenty-five. On one hundred acres, that's twenty-five hundred bushels. Soybeans were higher than a kite in 1972, over $3.00 a bushel. For our example, figure $3.00 a bushel or a gross of $7,500. After deducting costs on the crop, you might have a net of another $3,000. All of which means that on a two-hundred acre farm, you might clear close to the $10,000 above field costs (but not living costs, interest charges, or taxes).

A two-hundred acre farm is very small by modern commercial grain farm standards, a fact not many urban people realize. To them, two-hundred acres is a lot of space. To a grain farmer its little more than what he used to think of as the "back forty."

The reason I suggest barley instead of wheat for double-cropping in some areas is that barley matures earlier than wheat, allowing you to sow soybeans earlier after the grain is harvested. Also you can plant all the barley you want to and not have to worry about complying or not complying with government programs. And barley on good land should make yields of fifty-five bushels per acre or more—at least better than wheat. Though barley sells for less (around a dollar a bushel and a little higher right now) the larger yields make up for lower price to some extent.

I would not advise smaller, organic farmers to try to compete with any other kind of commercial grain operation. The odds are against you. The only occasions where I might change that advice are those where a special, high-priced market might open up for organically raised grain. This happened in the rice market in 1971. It could happen with other grains if consumers demanded the organic product. So far they haven't. And the organic rice market is back to normal too.

My grandfather always said, "The only good way to market

grain is through livestock"—that is by feeding it to livestock and selling it as meat on the hoof. However, for the smaller organic farmer, chances of making a good living solely from livestock farming, especially if you have to borrow money to get started, are very small. Don't expect to make it with chickens, beef cattle, or hogs unless you have a special, high-priced organic market. The chicken business is already dominated by very large operators, and there is no chance at all for a small farmer to make a living in chickens. The hog business is going the way of the chicken business. The price of hogs may be excellent right now, but over the long haul you have to produce an awful lot of hogs every year to make a living. Beef cattle demand the investments of the rich and the guts of a good gambler, or both, and is no place for the little guy without money. Raise beef for yourself, period.

Beef cattle demand the investments of the rich and the guts of a good gambler. If you can serve a specialized market for organic beef, the risk is lower.

You will hear about organic beef raisers who seem to be successful serving a specific market. If a good organic market for beef exists in some areas, a man might profitably take advantage of it for part of his income. But before you decide to go full time at this venture on the strength of what you have heard about such projects, *check out whether such small cattlemen are making their living from this project or whether they have another dependable source of income.*

Only in dairying is there a chance for a smaller farmer to make it, and even then, the chance is slim. There are still thirty-cow herds around which make a farmer a decent living. But high-quality, high-producing herds like that take many years to develop. A herd in which each cow averages twenty thousand pounds of milk is not beyond the realm of possi-

A small herd of top-flight cows can make a living for an organic homesteader. But milking twice a day, every day, means the homesteader has to be willing to really work.

blity. There are quite a few of them now. But they don't happen overnight.

So you know what kind of money I'm talking about, let us say you have a herd of thirty cows with a twenty thousand pound average and your milk price was, in round numbers, six cents a pound. That's $36,000, if my arithmetic is right. But that is not all of your gross income.

Each year your thirty cows will produce thirty calves, barring a fatality now and then. In a top herd with a twenty thousand pound average, these calves, whether as two-year-old bred heifers or as fresh cows, will bring excellent prices as replacement cows in other herds. Any cow that produces twenty-thousand pounds of milk is worth $2,000 if she is not old (twelve to fifteen years old is considered getting up in years); her offspring at weaning time, whether bull or heifer, may be worth $500 if the animal is a registered purebred. A little pencil pushing can show you the possibilities in dollars. If you have thirty head of cattle to sell every year, even at $500 each, that's another $15,000. Likely as not, a good herdsman will do better than that. And the man with the time to "prove out" a bull—keep a bull who proves his ability by siring daughters who produce significantly more milk than their mothers did—may have an animal that artificial breeding associations will pay $20,000 or more for—sometimes a lot more than that.

I have given you an example of dairying at its very best—a type of small farming one man can handle with family help and make a very good living. Your question should be: Well, if dairying is a gravy train, how come farmers are getting out of dairying in steady numbers?

The main reason I've already given: it takes a lifetime of love and skill to put together a good herd like that. Secondly, in dairying you can't say, "Well, if I do just half that good, I'll be all right." Won't work. Half that good is no good at all in dairying. Herd averages in this country are around twelve thousand pounds of milk per cow per year. At that rate, you don't make much money per cow. But the typical dairyman thinks it is easier to milk many cows at that production level than to try to build up or buy cows that produce at the twenty thousand pound level.

With only a small, one-man herd, the twelve thousand pound level won't make you a dime. Nor will your young stock bring you much more than the cost of feeding and keeping them.

At what level does a small herd become profitable? If you can't get production up to 17,500 pounds in three years, you better have another source of income.

A second reason why farmers aren't exactly rushing into the dairying business merits your serious thought if you are contemplating this venture. You can say it in one word: *Labor.* Dairying means milking cows twice a day *every* day. Successful small dairymen are all the stay-at-home types who find their joy in a steady routine—or in the near-at-hand breaks from it. And they don't stay out late at night. And they have to get up reasonably early every morning. And they can't call into the office and say they're sick. The cows *must* be milked.

That's enough to scare most Americans away in a hurry. But I want to make a point. Of the last five men I've met who seemed reasonably happy and at peace with the world, four of them were dairy farmers who told me with pride: "I haven't missed a milking in fifteen years."

Dairy farming—the small family dairy farm that raises all of the herd's feed on the place—comes close to being the perfect ecological farm. Feed—grain and roughage—is turned into milk and manure inside the cow. The manure is recycled on the land to produce more feed; the milk goes to market—the most nourishing and wholesome liquid young people can drink. Because of the cow herd's demand for lots of roughage in the form of hay and silage, the dairy farm keeps legume hays in rotation with other crops which aid the soil directly by building up nitrogen in it, and indirectly by lessening soil erosion.

If you decide to venture into dairying commercially, be sure to get some experience working on another *good* dairy farm first. You should have little trouble getting a job on a dairy farm, even part time, if you are conscientious. Dairymen have such a hard time finding good help they will welcome you with open arms. You may even use the job as a springboard to your own dairy business. I've known several "hired men" who received a calf or two each year as a part of their salary. They wound up owning a substantial share in the herd of cows and became business partners with their employers.

At any rate when you know about how a dairy farm should be run and have a place for the cows, the single most important step is the cows you buy to start your herd. A good dairyman in Ohio says it this way: "When you start out, take the money you were going to use to buy ten cows, and *buy two*." What he means is that you will be better off in the end by buying two high-priced cows who give a lot of milk, then ten average cows. If you have only $2,000 to spend, buy one or two cows with it and work at another job until those two cows build up to a paying herd or until you have more money to buy more good cows. Don't buy a bunch of cheap cows just so you can get into dairying full time right away. It won't work.

As a small dairyman, to make a profit you will have to substitute labor (yours) for capital at every opportunity. For instance, if you do not have a silo, it would be "easier" to have one built and fill it mechanically each year with silage. But it may be cheaper for you to forgo the silo and make more hay, a job that requires more hand labor than silage, but less outlay of capital.

However, no matter what kind of feeding program you

decide to use, you will have to avoid buying new or unnecessary equipment when you have only a thirty or forty cow herd. You do not need to buy new tractors; second-hand smaller models will do just fine. You cannot afford automatic loaders and unloaders nor most of the new push-button gadgets. Not with a small herd. Don't let salesmen fool you with that "your time is too valuable to spend it doing. . . ." You are trying to beat the odds by following a sort of old-fashioned notion that a small farm can be profitable. Therefore, use some old-fashioned philosophy which says that if you have to borrow money to buy new equipment, something's wrong. Borrow money only to buy good cows or good land—and don't do that very often.

The kind of expansion you are aiming for doesn't cost money. You intend to keep your cow herd fairly constant, say at thirty cows. But you intend by good care, good feed management, and good genetics, to raise the production level of those thirty cows every year. If you get up to a twenty thousand pound herd average, you will be very much a success. And the challenge can be a most exciting one to accept. Furthermore, there is no limit yet established as to just how much milk a cow can give—some champions have produced more than thirty thousand pounds in one lactation.

This kind of expansion costs you time, but not much cash. The costs and labor in keeping thirty poor cows is almost as great as with thirty good cows. And thirty good cows often make a farmer more profit than one hundred mediocre cows do for his neighbor. To repeat the secret for the beginner. Build up a *good herd* with the speed and only the speed that your money allows. Keep your off-farm job in the meantime.

Actually that advice holds true generally for any kind of farming you decide to take up commercially. But an awful lot of farmers fail because they don't heed it.

Chapter Eleven

Blending Technologies Old and New

I am no sentimental or nostalgic dreamer when there's work to be done. If you want to use a crosscut saw, more power to you; I'll stick with a chain saw and hope that some day manufacturers will muffle the noisy things.

On the other hand, I happen to know that some of the tools of the past are better than the new gadgets that have taken their places. For instance, the chain saw is my selection for cutting down a tree. But for triming off side branches, a good sharp ax or bow saw beats the motorized monster all hollow. Notice I said a *good sharp* one. One reason modern man is loathe to use hand tools is because there's no one around anymore who can sharpen things properly.

For small- to medium-sized gardens, the traditional hand tools—hoe, rake, and spade—remain far more practical than larger, mechanically powered tools. Most of the cultivating attachments for rotary tillers and small garden tractors are more bother than they are worth (but not the rotary tiller itself!). The hand-pushed cultivators should be used in place of the hoe on slightly larger gardens. You can cultivate a quarter acre of garden with a hand-pushed weeder easily. I

Agriculture is increasingly mechanized, but hand implements still have their place. And their place is on the small organic homestead.

have no trouble talking my wife into helping—she gets exercise and sun-tan pushing the thing when I get tired. The cultivator makes no noise, always starts, never breaks down (a tree fell on mine once, but I patched it up quickly enough), needs no gasoline, can be controlled easily to avoid plowing out vegetables—and mine is at least fifty years old.I have a personal vendetta against horses, twice having nearly been killed by the blamed things—once when I was thrown and once when I was the unwilling driver of a team of runaways. However, if you need more power than your muscles in small farming, you might want to give a thought to horses rather than a tractor. With horses, as with everything else, there are good ones and there are poor ones (I have always been stuck with the latter). A good team (whether horses or mules) can be an asset to your organic homestead, very much in keeping with the spirit of organics.

Horses do not produce carbon monoxide; they fit into the farm's natural recycling system in a very economical way.

If you don't believe it, visit the stricter Amish farmers in Lancaster County, Pennsylvania and watch them work with horses. Then go to the bank and find out which farmers have the most money stashed away. Likely as not, it'll be the man who farms with horses.

However, if you don't like horses much, like me, or if you don't have the foggiest notion of how to harness a team and don't want to learn, a smaller used tractor can be economical,

The horse can do many jobs better than a tractor. Moreover, unlike a tractor, a horse contributes to the production of its feed and its replacement. Worth considering for a small spread.

too, as I mentioned earlier. Just don't automatically nix the idea that horses aren't still practical on small farms.

In the late 1950s, when work horses were already beginning to become a curiosity on most farms, the farmer I worked for in Minnesota kept three teams, along with a couple of tractors. Once you got used to the bother of harnessing and unharnessing the "critturs," their advantages over tractors in certain operations became apparent. Horses can pull a wagon full of fence posts over rough terrain where tractors can't go. If you are picking up hay bales in the field, you can start and stop a good team with your voice while you load bales from the ground. With a tractor, you either must have another worker to drive, or keep climbing on and off the seat to move to the next bale.

It goes without saying that if you have cattle scattered through wooded hills and ravines, riding horses are about the only way to round them up. It also goes without saying that if you have perhaps only two acres of each crop you need to cultivate, a horse or mule will handle the job far more cheaply than a machine. And if the crop is trellised or grown on stakes you can keep the rows closer together for horse cultivation than for tractor cultivation.

The horse is but one example, and a minor one at that, of the organic homesteader's ceaseless search for alternate sources of power that are ecologically sane and available for his independent use. Old ways of using wind power, water power, and sun power are now being re-examined, not only as substitutes for polluting power sources, but because we face a power shortage, especially of electricity, which conventional production methods may not be able to relieve. So far, atomic energy is being touted as the answer to this dilemma, but many scientists fear the accumulated fallout from atomic reaction could be more dangerous than present pollution problems. As a result, it is no longer just crackpots like myself wondering aloud if solar energy is not a far more feasible answer to power needs in the long run. Nor am I the only one who thinks that wind generators could not be made more efficient. Buckminster Fuller, the famous inventor, is working on a model of such a generator—it's supposed to be eighty-five percent efficient.

Solar energy as a direct source of power is, of course, an old idea. Inventors over a century ago constructed sunlight "traps" to collect enough heat to run a steam engine. But the process has always been inefficient (which means more costly than conventional sources).

However, in southern climates, houses can be constructed to gather enough sunlight to warm water sufficiently to heat the structure, even overnight. The hot water "stores" the sunlight, so to speak, until the sun shines again. In spaceships beyond the earth's atmosphere, sunlight becomes a steady and reliable power source for batteries. All science really has to do is come up with a battery that can store power between relatively long periods when the sun doesn't shine, and every home can become it's own power plant—without pollution. If you have ever set a piece of paper on fire by training sunlight through a twenty-five-cent magnifying glass, you don't find it too hard to believe that when man devotes as much energy to solar power as he does to wars, he will lick the problem.

Scientists like William Heronemus of the University of Massachusetts are advocating more research into solar energy as a safer power source than atomic reactors. But few people believe that solar energy could supply all the electric power needed. This should not discourage the homesteader who wants power only for his own use, and not in quantities the typical home uses. He has built a well-insulated house, he has planted trees around the house (or built the house amid the trees), and to keep cool he will use a fan rather than an air-conditioner anyway; in winter he may need only a supplementary source of heat rather than an "all-electric" home. He will not waste time or money on ridiculous gadgets like electric toothbrushes. He doesn't leave the refrigerator door open any longer than necessary, turns off lights when not in use, dries clothes outside in nice weather, may not have a dishwasher at all, keeps the house at a healthy and comfortable 70 degrees rather than 80, and, in short, practices all kinds of economies which result in using only half as much electricity as the "typical" home. What I am saying is that just as families who rely on cisterns for their water supply learn to use water frugally, so too can smart homesteaders live comfortably on a power unit producing less electricity than the scientists now think is necessary. Few scientists advocate cisterns either, for the same reason, though, as I will show a little later, cisterns can still be a very satisfactory source of water.

We already use solar energy directly. That's what a greenhouse does with only one layer of glass. In Israel the greenhouse concept using more than one layer of glass has resulted in "solar trap" devices which heat houses adequately. Farmers in this country have learned how to dry hay artificially by circulating heat collected under the black roofs of their hay barns. Michigan State University has an excellent set of plans for building a shed to dry hay in this manner.

You use sun power to dry fruit, spreading peach, apple, apricot, plum, and raisin grapes in thin layers on a surface exposed to direct sunlight. Cover the fruit with a thin muslin cloth to protect it from bugs, and the sun does the rest. When you hang a smoked ham under an attic roof, it is heat from the sun that drives the salt into the meat. You can keep water from freezing completely across the surface of a large horse trough or storage tank by covering it partially or totally with a layer of boards, four inches of sawdust, and a covering of black tarpaper. Between the sawdust and the heat absorbing tarpaper, it takes a very cold night to freeze a float valve underneath.

Wind power is another energy capable of producing electricity. City people are surprised, especially younger ones, to learn that wind machines were used extensively in places like South Dakota forty years ago. Users will tell you the machines generated enough power "to get by on" most of the time. A good friend of mine says wind generators are still used on some farms on the plains for supplementary electricity. Wind generators are sold by Wincharger Corporation, which has been sold to another company, Dyna Technology, Inc., a subsidiary of Southern Gulf Utilities, Inc. Address: East Seventh At Division, P.O. Box 3263, Sioux City, Iowa 51102. Only a twelve volt charger was being manufactured when I contacted the company. The thirty-two volt and hundred-ten volt were discontinued in the middle sixties.

The windmill can harness cost-free energy to pump water. Windmills are still in use in many parts of the country. Your homestead should have one.

The windmill is one "old" tool that deserves to be used on farms forever. Once installed, a windmill pumps your water at a very small cost. A little grease and oil will keep it running a lifetime. Or two. Electric pumps have replaced most windmills except in field pastures far from electrical outlets, and that is unfortunate. If one *must* have an electric pump, he ought at least keep the windmill in place and ready for use. Today, many farms are in the embarassing situation of being entirely dependent on electricity to get water out of the ground. In case of power shortage, no water.

Old windmills always had wooden storage tanks nearby. When the wind blew, the mill pumped water into the storage tank so there was always water available when the wind did not blow. Clever farmers who picked their homestead headquarters with a sharp eye would put the windmill and storage tank on a hill above house and barn. The water could then be piped by gravity to all buildings. Sometimes a windmill itself was equipped with a small storage tank toward the top, again taking advantage of gravity.

Gravity is the force that makes water-powered mills and generators work. Falling water is energy in the raw, and modern homesteaders should be as aware of the possibilites of harnessing it as their ancestors were. Make it a point on your next vacation to visit one of the still-operating water-powered gristmills and study it. You'll get an idea or two.

The gristmill I am most familiar with is at Spring Mill State Park in southern Indiana, where both an overshot wheel for grinding grain and a tub wheel to power a sawmill have been renovated and made operative. Water is diverted from a spring-fed hill stream some distance behind and above the mill. The stream has been dammed, forcing water to travel by sluice through to the mill wheel. A pipe off the sluice sends water directly down against the horizontal tub wheel that powers the sawmill. After you watch that old "up-and-down" saw walk effortlessly through large logs and turn them into boards with no more help from man than the flicking of a few levers, then you know we have not really progressed so far since the "old days" after all.

I have visited one privately owned gristmill, newly restored, in which the huge water wheel, an undershot model powered by water channelled directly out of a nearby river is used to run a modern generator that produces enough current for all one-hundred-ten-volt outlets in the mill. Renovation of the mill was most costly and beyond practicality for most of us, but the point is, electricity doesn't have to come from Con Ed.

When electricity first came into general use, many farmers went about manufacturing their own as a matter of course. All you needed was a small stream capable of producing at least one horsepower to operate battery-charged light plants. With a four-horsepower stream, you could operate a lighting plant without a storage battery by using a turbine with a generator which would deliver a constant voltage.

How do you tell if your stream will develop enough horsepower to generate usable electricity? First you find out how many gallons of water flow past a given point per minute. The old way to do that was to put up a temporary board dam with a weir, or notch, cut into the middle of it. The amount of water flowing through the weir was determined with this formula: three times the width of the weir times the depth of the water flowing over the lip of the weir times the square root of that depth. A weir otch twenty-five inches wide with a water flow over it four inches deep would, if my mathematics is correct, denote a six hundred gallon per minute flow.

To determine horsepower roughly, multiply the gallons per minute flow times the height of your dam, or the number of feet the water will fall from the top of the dam breast to the water surface below. In the case of a six-hundred-gallon-per-minute flow and a seven-foot dam (head), horsepower potential would be a little over one—following the formula that says: rate of flow times height divided by four thousand equals hp. A one-thousand-gallon-per-minute flow and an eight-foot dam would develop two horsepower. A one-thousand-six-hundred-gallon-per-minute flow and a ten-foot dam would develop four horsepower.

Another method to measure the horsepower potential of a stream is the float method. Let a piece of wood float down a straight stretch of the stream, and measure how far it floats in a minute. The float must be weighted—that is, have a weight tied to it, hanging down in the water so that surface currents or wind will not push the float along faster than the water is actually flowing.

Let's say the float moves twenty-five feet in a minute. That's your velocity. Multiply that by .75 to get average stream flow, giving you true velocity of 18.75 feet per minute. Then you must make a calculation as to the average width and depth of the stream. Multiply that times velocity to give you cubic feet of water per minute. If average width is six feet and average depth 3.5 feet, with velocity at 18.75 feet per minute and a nine foot dam, you can then determine horsepower with the formula: Horsepower equals the weight in pounds of a cubic foot of water (62.5), times the flow in cubic feet per minute, times the fall, divided by a constant, 33,000. You multiply that quotient by .80 because falling-water energy is reckoned only eighty percent efficient. So, to determine the horsepower of the hypothetical stream, you multiply 6 (the average width) times 3.5 (the average depth) times 18.75 (the velocity) to get the flow in cubic feet per minute. Multiply that by 62.5 (the weight of a cubic foot of water) and by 9 (the fall). Divide by 33,000 for a theoretical horsepower rating of 6.7. Multiply by .80 for the practical horsepower rating of 5.36.

You can install the water wheel at the dam or run the water

by pipe or sluice to a wheel next to a building which houses a generator and storage batteries. If your fall is under five feet, an undershot wheel (one where the water propels it by pushing against the bottom of the wheel) is your only chance. If fall is between five feet and twelve feet, an overshot wheel is better. (With an overshot wheel, the water runs over the top of the wheel, its weight moving the wheel on around.) If you have twenty feet or more of fall, especially if the amount of water is small, as with a small stream on a mountainside, an impulse wheel is best. Here, the water is dammed up, then carried down hill in a pipe. A nozzle at the end directs the water on the wheel under high pressure.

By such methods were generators run in the past. Storage batteries sat in rows in the farmer's "powerhouse" and old farm papers carried many articles on proper care of these batteries. Inevitably, it became more economical and a lot less bother for electric companies to generate electricity and distribute it by wire to paying customers. Further work on developing better and more efficent types of storage batteries was simply not done. We're still looking today.

Unfortunately, most of the older DC low voltage electrical equipment used by our forefathers is hard to find these days. You will have to know more about electricity than I do to solve that problem. But where there's will, there's ways.

I would not like to be without electricity of some kind for very long, but you can get along without it in your barn without too much inconvenience. Every homesteader should have a lantern or two and some coal oil or kerosene to fuel these old-fashioned "flashlights." I have two, and use them both in the barn at night and in the garden, where I sometimes work after dark. A lantern is convenient and easy to use. Kerosene lamps in the house will do in a pinch, too. I can remember from my childhood living very well without electricity—we did not know we were living primitively. We went to bed earlier and got up earlier, too, which would be an excellent idea for more people to practice now.

The fireplace, one of man's oldest inventions, refuses to become obsolete. Most people build fireplaces for nostalgic or artistic reasons, but you can still heat a room and cook meals with one when the electricity goes off. Advancements in outdoor camping equipment and barbecue cooking have led to some pretty sophisicated and convenient utensils. The homesteader should be aware that you can cook outside with charcoal today almost as easily as you can cook inside with electricity. Some of those fancier barbecue stoves could be used inside, too, if fitted with a smoke pipe to the outside.

Information on how to make charcoal is easily available from the library. I doubt if many homesteaders would want to bother, charcoal being reasonable to buy, but making it is not difficult once you understand the principal involved.

Don't laugh at wood stoves, either. My Maine friend I have mentioned before has stuck with his for over fifty years, and it gets pretty cold in his country. He has given in to progress to the extent that he bought his wife a gas range to cook on, but the old black wood stove dominates the little house and provides most of the heat except in the dead of winter. Then he may stoke up the furnace. He doesn't like to, but the rest of the family finally insisted on some kind of central heating. As far as he is concerned, the stove is heat enough. He likes to pull down the oven door and rest his stocking feet there to catch the warmth from the oven. As long as he can cut his own wood, he says, that stove is by far the most efficient heater ever invented.

In the country you should not scoff at the outdoor privy. The privy has more ecological sanity to it than cesspools and sewers. But I was almost flabbergasted to visit an organic homesteader who had just built a *new* privy.

I've used enough privies to know a good one when I see it.

The privy is often regarded as a picturesque relic of the old days. But it's ecologically sane, relatively inexpensive, and quite appropriate on the modern homestead.

This one is built into the side of a hill, like a split level house. You enter the three-holer at ground level on the hill top side. The waste drops through to a deep pit on the bottom floor. On the lower side of the hill, a door opens to the pit. Inside there is room enough to transfer the waste from one pile to another and compost it with other organic materials like straw, peat moss, and lime. There is very little unpleasant odor. The finished compost is used to fertilize ornamental plants, not food plants.

The "bathroom" itself is painted a gleaming white, screened against flies, and altogether as clean as a hospital room. My only complaint—a slight disappointment in finding toilet paper instead of the traditional old Sears catalog.

I asked the owner why he built the privy. "Well, at least I know *I'm* not putting more bilge in the river from a sewer or a cesspool," he answered. "But it's more than that. Modern man seems to think he has evolved beyond his animality into a sterile world of sugar and spice and everything nice. We can ignore the evils of pollution because we have made it easy to sweep our wastes under the rug, so to speak. Just flush the toilet and down it goes. Disappears. What we don't see, we can pretend doesn't exist. When you live with a privy you are forced to face reality. The flush toilet is the porcelain throne in our ivory towers."

Manure is getting to be more respectable these days. Scientists have been holding seminars in Washington on the subject—especially as to the affect of animal manures on ground water and streams. Out of the brainstorming have come some interesting ideas.

The first of these is another look at manure as a source of methane gas for fuel. Nearly everyone knows now about the Englishman Harold Bate, who ran his car on methane from poultry manure. But German, French, and Indian farmers have also been busy building more practical home digesters which convert manures and other organic wastes to methane gas. The digester can be any large tank with an airtight lid. Attach a pipe and valve and you have "bottled gas" that, properly filtered, can be used just about as you would propane. Best source of information are two booklets from Ram Bux Singh, Gobar Gas Research Station, Ajitmal, Etawah (U.P.), India. Cost: $5.

Researchers have lately learned that manures can be converted into oil, though the process is not economical yet. Manure *is* being used to make tiles for interior walls and blocks for construction purposes.

I had almost decided that cisterns were becoming obsolete and that I was becoming something less than wise in continuing to praise this relic of a bygone era. Then to my surprise, I visited a homesteader in southern Indiana who had just built a cistern. What made this particularly newsworthy was the fact that the man who built it is an engineer who works as a geologist specializing in *water.* "Of course cisterns are practical," he says. "The reason more people don't have them is because they don't know any better, and the water companies for sure aren't going to lose business by informing them any differently."

This fellow's cistern is a big one—holds sixteen thousand gallons. "I have five children. We use a lot of water. My wife has a modern washing machine, which as you know is not an efficient user of water. Yet we have never run out of water."

The top of this cistern is also the cement floor of the porch. The water is piped into the basement *through a filter of sand.* "The water just sort of seeps through the sand filter, which is about six feet by six by eight in size. I've got two well points sticking into the sand and the water is pulled through into a tank in the basement." He has a chlorinator rigged up to the water tank and uses regular laundry bleach to chlorinate the water.

This homesteader ignores all the old cistern rules that I thought were gospel. He has no charcoal filters for the water to run through when it comes off the roof; he saves the water during May, June, July, and August—times we were always told were not for putting water in the cistern; he does not even allow the roof to wash off before diverting rain water into the cistern. "Between the sand filter and the chlorinator I've got water as good as anybody's water, and I know. That's my business," he says.

Our cistern always used to run dry in the summer because it wasn't big enough. Then we'd have to buy water hauled in by the truckload. A lot of people in areas where wells are hard to find still buy much of their water this way. They wouldn't have to.

The man with the new cistern has a ranch style house with lots of roof area—he planned it that way, thinking of a cistern. People used to figure that a family of four needed twenty gallons of water a day. You can still get by on that—five gallons per person—but you better double it or triple it for fastidious Americans who think their birthright is a tubful of water every day. (When I was a kid we were allowed three inches in the tub except on our birthdays. Then we could fill 'er up.) At fifty gallons per day for a family of four, that's a little over eighteen thousand gallons a year. *The old rule is that your cistern should hold at least half of your yearly need.*

You can determine how much water comes off your roof in a given rain with this formula: Multiply the square feet area of roof by the inches of rainfall and divide by 1.6. Thus a thirty-foot by forty-foot roof with an inch and a half of rain puts 1,125 gallons of water in the cistern.

Cisterns can be made of poured concrete using forms or, if you know a foolproof way to waterproof them, of cement blocks. A round cistern, eight feet deep by ten feet in diameter holds 4,700 gallons. A rectangular cistern eight feet by eight feet by ten feet holds 3,840. All downspout entrances to underground pipes leading to the cistern must be screened to keep mice and bugs out. If you buy an old country house that already has a cistern, drain it, scrub down the walls, and patch any leaks before trying to use it.

With a little extra work, a cistern pays off. First, you don't need a water softener. There is nothing softer than rainwater. Modern man misses the delights of a bath in it, and modern woman misses the delights of prettier hair washed in it. With rain water you use only a smidgin of the amount of soap necessary in harder water, saving you money. Clothes clean faster in many cases with only half the fabric softeners, bleaches, and stain removers that seem so necessary. Moreover, far less rusting of pipes and fixtures occurs with rainwater as compared to hard water. Finally, many horticulturists insist that rainwater makes plants grow better than well water or "city" water does. If you don't have a cistern, you ought to at least have a rainbarrel under one of your downspouts if you grow indoor plants.

A good well with a windmill above it is the easiest, safest, and most reliable source of homestead water, and I personally would not buy a small farm without one on it, or where there was reasonable doubt that one could be dug. But many small farmers in the past have "made do" by utilizing pond water for home use. The most economical way to do this is as follows.

Dig a small pond below the main pond—needs to be only big enough to hold six-weeks to three-months supply of water. Water runs by gravity through a pipe into the small pond. In the small "settling pond," sprinkle around the edges a mixture of one part lime and two parts alum at the rate of a half pound per one thousand gallons of water. The treatment helps the water to settle out clear. At the lower end of the settling pond, form a sand bar about six feet by two feet and two feet deep. This sand acts as a filter, just like in the cistern I mentioned. Under the sand, place a perforated pipe into which the water will seep. Once out of the sand filter, the pipe is solid and takes the water on to the pumphouse. By that time, the water may be clean and pure enough for use, but most states require that you chlorinate the water—and you should whether required or not.

A good spring is worth—well, I'd pay a thousand dollars for one, and my grandpaw (who was richer than I am) said a good, cold one was worth $2,000 in his day. Sometimes you don't have a ready-made spring, but along the lower side of a hill you'll see a sort of seepage of water, just enough to keep the ground muddy in the area. You may be able to turn this seep into a spring. Here's how. Run two fifteen-to-twenty-foot drain tile lines above the seep area, running across the slope. The individual tiles should be placed with a quarter-inch crack between them and about two feet deep in the ground. The two lines, like the two arms of a "Y," should join a main tile line running down the slope. The junction of the two branches and the main should be located just above the main seep spot. Just below the junction, build a cement collar about two feet high, four feet wide and two inches thick around the main tile. That collar holds back seep water so that it will more easily drain into the tile rather than on down the hill through the ground. The main tile line runs to a cement tank at the foot of the hill. Most likely you will get only a small trickle of water into your tank, but a steady trickle amounts to quite a bit of water. The concrete wall of the tank should be at least three inches thick to prevent cracking if the water should freeze.

Good springs located above the level of a farmstead are

Don't sniff at an old barn, if you're lucky enough to have one. It's housing for livestock or a source of good lumber. Some people with skill and determination (and usually money) have even converted them into homes.

utilized extensively where practical to supply all the water for household and livestock. The water flows through all the buildings by gravity. Many eastern Pennsylvania farms are still watered in this way—one I know uses a spring a quarter mile away, the water being carried to the buildings by pipe installed fifty years ago.

A springhouse is not as good a cooler as a refrigerator, but can be an excellent place to store apples and vegetables in both cold and hot weather. If you are lucky enough to own a farm with a springhouse, count your blessings.

Don't discount any old buildings you may inherit when you purchase a farm. Often broken-down barns contain good lumber you can salvage for new buildings. The price of lumber being what it is, that old wood can amount to considerable savings.

Information on how to build low-cost wood houses is available from the Department of Agriculture. Send for Agricultural Handbook Number 364, "Low Cost Wood Homes for Rural America—Construction Manual," Superintendent of Documents; U. S. Government Printing Office, Washington, D.C. 20402. It costs a buck. The handbook describes eleven designs for simple rural homes. Full scale plans for many of them are available at $1.50 each.

From tree limbs too small in diameter for any other practical use, you can fashion rustic outdoor furniture, benches, stools, feed racks for animals, and gates. Farmers' Bulletin Number 2104, same address as above, tells you how to treat such wood so it will last a long time against rotting and bugs.

Having lived three years in a log cabin, I can vouch for their comfort and practicality. But log cabins are wasteful of wood and therefore not necessarily cheap if you have to buy the logs. However if you have your own forestland and a good chain saw, you can put up an adequate building very economically.

Many books and articles have been published lately on how to build log cabins. The *Foxfire Book*, published by Doubleday, is one of the better sources of information. *Organic Gardening and Farming* is another. Only advice I would add is to build a small log shed first, for practice. Use it for a chicken coop or calf shed. By that time you will know how to build a log cabin, and if you want to.

Best kind of log for cabins? Yellow poplar, if you have it. Yellow poplar (tulip tree) grows straight as an arrow and very tall in dense forests. That's why the pioneers liked them for cabin walls, too. But I'd use any kind of hardwood I could get a good log out of, except walnut or cherry which are much too expensive to use for cabins. But I wouldn't be afraid to use pine, a softwood, either.

One thing to be sure of: Get a stone or cement foundation under the logs—high enough so that the bottom log is a foot off the ground. And peel the bark off the logs before using them.

Log cabins are extremely well-insulated even without adding insulation—if the logs lay well against each other and need only a minimum of clay to fill in the cracks. There's something honest about a log cabin, and living in one you somehow feel more at one with the universe and your land. Your wife may eventually persuade you to cover the inner walls with some god-forsaken factory-made panelling, but put her off as long as you can. The honesty of bare logs has more than economy to recommend it.

The soil on your homestead can be useful in other ways than for growing crops. You can build a house out of it. Building bricks made of soil and straw—one part straw to five parts mud—make adobe houses that in arid and semi-arid regions will last nearly forever. In humid regions, you can make stabilized earth blocks using a little cement or asphalt—one part cement to twelve parts soil. You can mix the mud for your blocks by hand or in a pug mill, and you can use wooden forms or a block press, several kinds of which are on the market, to mold them.

You have to experiment to find the kind of mud that makes a good block for you. Start with a sandy clay loam. Make a sample block, and let it dry. If it warps or cracks when dried, there is too much clay in the soil you are using. Add some sand. Make another block. If it crumbles on drying, too much sand. Add clay.

Protect the blocks from frost until they are cured. Don't try to make blocks in freezing or rainy weather.

The advantages of a house made of dirt are many: the building material costs nothing; the massive walls are strong, durable, fire-resistant and exceedingly well-insulated (soil is a great insulator). The disadvantage is that blocks will deteriorate if exposed to water over a long period of time. Also, there is a lot of labor involved in making them.

To protect walls made of soil in humid, rainy regions, you can waterproof them. Mud or stucco plasters work well for this. Mud plasters must be painted to withstand weather. Stucco can be either lime stucco or cement stucco, the latter being much more durable. Be sure to allow the walls to dry and settle for two months after they are up before stuccoing.

Information on building with soil can be obtained from the Superintendent of Documents in Washington, same address as mentioned previously. Ask for Leaflet Number 535: "Building With Adobe and Stabilized Earth Blocks."

Clay and sand have other uses that might come in handy for you. Boxes of dry sand make good storage receptacles for vegetables rescued from the garden for winter. Carrots, potatoes—any root crop will keep fairly well this way.

Clay can be turned into useful dishes, mugs, and pitchers if you have a kiln or know someone who does. To get proper material we Logsdons go to our favorite clay bank along a nearby creek and lug back as much as our backs will let us. The clay then sits by the furnace in a large tray until it is dry. Then I pound it to powder, strain it through a screen—regular window screen works, but a finer mesh would be better. I mix the powdered clay with water—about equal parts of each—and allow the mixture to slake. All that means is that you let the clay–water sit quietly until the clay settles out to the bottom of the container. Then pour the water off. After awhile, the clay will firm to a good working texture. When it does, put it into plastic sacks, tie the sacks shut so no air can get in, and store them in a cool place. I have stored clay this way for over a year.

My skill with clay if very limited—I confine myself to the slab and coil methods of construction. I know just enough to tell you that you don't need a potter's wheel to make many things.

Rocks and stones on your homestead are free for your use and as New England farmers learned, can be very useful indeed. Rocks make good fences, right? Also barns, houses, patios, walks, and missiles for throwing at strange dogs.

I have built stone walls—two little ones—and my hat is off to artisans who have mastered this craft. To lay up a stone fence or wall without mortar is a skill I have failed to understand, much less master. To be able to do it with rocks that are not necessarily flat is a plain miracle to me.

If you decide to build walls of rock, best to go to the quarry and hand-pick the flatter, more squarish rocks—all about the same size. If you plan an eight-inch-thick wall, all your rocks should approximate that measurement.

If you cannot get your stones yourself, instruct the quarry truck driver as to the kind of stones you want. Most quarrymen are used to the quirks of both amateur and professional stone wall builders and will try to bring you flatter, squarish rocks that lay up better. They can also help you figure the

You don't have to be a stonemason to erect stone walls, as this Logsdon-built wall demonstrates. Stone walls serve many functional and decorative purposes, and the basic material is sometimes available right on the site.

approximate tonnage of rocks you will need if you know how long, high and wide you want your wall to be.

Probably the easiest way to build a rock wall is the method used by bygone farmers in parts of Minnesota where rocks are plentiful. They simply built a form for the walls of their barns, poured in some concrete, and while it was still wet dropped in the rocks. Then more cement and more rocks until the wall reached the top of the form. After the wall dried, the forms were taken down and used for another wall. The finished wall was not quite as attractive as one where the rocks are laid up by hand, but it still effected a certain beauty. And the wall will last forever.

There are other imaginative uses of the materials and energies of nature around you, which publications like *Organic Gardening and Farming* or the *Mother Earth News* continually describe. My purpose is to aim you in the right direction. We live in a world in which everyone under the age of forty (at least) has been taught that what you need you must buy. Advertising has made buying a reflex action, if not a matter of pride and status. The homesteader discovers how to use the near-at-hand that has a price tag of labor-only on it. Upon that rests the success of the small homestead. Once you realize that your head and hands are more important than your pocketbook, the rest is possible, if not easy.

Chapter Twelve

What It's All About

You may conclude by now that the small organic homestead is too much work and not enough play. That depends on how you define the terms. For the homesteader, work is what you must do for a living, but would rather not. Everything else is play.

I'm sure if I had to cultivate gardens ten hours a day every day for someone else, I'd think of it as work. But the beauty of the organic homestead is that "work" is self-willed, not commanded from on high or dictated by economic necessity. "Work" becomes creative, individualistic, done out of love, not someone else's sense of duty.

But beyond the activities that might be termed play–work or work–play, the successful homestead provides opportunity for pursuits of a purely recreational nature. If your home schedule does not provide time for simple reverie in a fence corner you've failed somewhere. If a hammock—well-used—is not among the accessories of your homestead, you're doing something wrong.

If you have been brought up on a farm, you already know the many ways country people have to amuse themselves.

And if you are originally from the city, you may know some hobbies new to country people. Here's a list—far from complete—of both kinds:

1. Hunting Indian relics. I've found arrowheads in almost every locality I've looked for them. That includes Ohio, Indiana, Kentucky, Minnesota, and Pennsylvania. I find them in streams, along stream banks, and especially in fields bare of vegetation, such as corn fields after harvest or before spring plowing.
2. Bottle hunting in old trash piles and around abandoned farmsteads.
3. Collecting butterflies—or any bugs for that matter.
4. Birdwatching.
5. Ice skating, sledding, tobogganing. For delightful holiday fun, hitch a team of horses to a big wagon sled (if you can find one), and take a gang of friends for a ride over snowy hills.
6. Hayrides. And barn dances.
7. Swimming and fishing.
8. Hunting and trapping.
9. Whittling. Good for the soul.
10. Outdoor photography.
11. Landscape painting, if you're up to it.
12. Gathering nuts, wild persimmons, pawpaws, or wild plums. Making cookies using some or all of these ingredients.
13. Picking wild grapes and making wine.
14. Starting a rock collection.
15. Building a canoe.
16. Exploring abandoned farmsteads (with permission). In the west, go "ghost-towning." That's what ranchers call visiting *real* ghost towns. In some areas, as southern Indiana, spelunking is a typical rural hobby, too. Lots of caves there.
17. Horseback riding.
18. Hunting antique insulators along telephone lines.
19. Collecting old barbed wire. Strands of about eighteen inches are bought and traded among collectors.

20. Hunting driftwood, dried weeds, and so on for table decorations.

21. Hiking, picnicking, camping out. You can pursue such pastimes right on your own farm. Some farmers build vacation cabins back in their woodlots or along their ponds and take restful vacations without leaving home.

22. Archery; target shooting.

23. Herb collecting in the wild.

24. Bicycling—a lot safer on country roads.

25. Recording folksongs, folktales, and other forms of rural, oral folklore.

The list could go on and on, depending upon your own interests. The point is, there is never a dull moment in the country. And that's why you have chosen to live there.

But the organic homestead means something deeper than either the nobility of work or the pleasantness of leisure. What it must provide—if the homestead is to have true success—is a shrine to tranquility, an island of calm sanity to which you can retreat each day from the hectic outside world.

And what is tranquillity?

Most visitors to our home become alarmed when we proudly point out a huge gray hornets' nest hanging from the porch ceiling uncomfortably close to the entrance to the house. But when my sister visited us (she's a country woman who knows a thing or two about hornets and such like), she made a different observation, which I consider the best compliment I've ever received. "You must have a peaceful environment around your home," she mused, staring at the nest, "or those hornets wouldn't have built a nest on your porch. They know there is not much fear or strife here."

I would like to believe her observation is true. We certainly try hard enough to make it true. At least I can say the hornets have never been alarmed; we have never given them cause for alarm. Sometimes when we ring the dinner bell which is just a few feet from their nest, they become excited—or did at first. But they seem to have gotten used to that, too. We can stand right beneath the nest—I have climbed up and stared right into the entrance—and the winged stingers pay no attention. They do not fear us, because they know we do not fear them. We both know there is no good reason why we cannot share the porch.

Our hornets are a very small but significant example of the basic philosophy of the organic homesteader: accomodate yourself to nature whenever possible, don't dominate nature when you don't have to.

I am not suggesting that nature is free of strife and fear, or that the natural way is never the violent way. There is a kind of violence that forms the fiber of nature: all life feeds upon other life.

But man has made a science of violence. Evolving into a world where he had at least three strikes against him from birth (man is born the most helpless of all animals), the human animal, fearing for survival, learned *cunning* violence. He learned overkill.

The birds and the animals battle for territorial rights and for mates. Man, supposedly being smarter, recognized that such violence could be avoided by enforcing some kind of Law. But the same mind that conceives of law also invents lethal weapons that kill on a scale far beyond the natural limits of natural violence.

A culture built on fear and violence cannot acquire a true morality. Without peace with nature, there can be no tranquillity among human beings. Men who can for economic gain bombard a forest or a field with a poison that can indiscriminantly kill the insect life therein can easily be brainwashed into believing there is a necessity to drop bombs on other people. The man who will shoot wild animals for no reason other than to prove his skill at aiming a gun can readily be trained to shoot other people. The man who brags that he

The reality of the organic homestead may fall short of the dream; life in the country can be beautiful, but perfection doesn't exist. But there are real homesteads that come close.

has worn out three farms in his lifetime is brother to the man who brags he has worn out three women in his lifetime.

This is why it becomes important to the organic homesteader what kind of fertilizer he uses on his beans. This is why he will risk ridicule of the worldly wise to ask: "What else will your new product do besides make profits for everyone?"

Recently a stone-age tribe of "uncivilized" people was discovered in the Phillipines. Surprisingly, these people are happy, content, peaceful—they live a completely organic life, in tune with their natural environment. They need protection only from the civilized people around them, say anthropologists, which is certainly the severest criticism civilized man has ever received.

The organic homestead is both a way back toward the innocence of these primitives and a way forward to a more intelligent use of what civilized man has learned. Man can live and let live to a much greater degree than he ever could before. Cunning violence, the violence of overkill, is obsolete. We can teach ourselves to walk our way through existence with a softer step and a gentler hand. Conviction begins on the organic homestead.

And what if there were millions of organic homesteads? A nation of them? Jefferson had such a utopian dream, so I guess its all right if I dream that way, too.

Bibliography

Other Helpful Readings

I think everyone who has any love for the land and for poetry should read Wendell Berry's *Farming: A Handbook* (Harcourt, Brace, Jovanovich, Inc., New York, 1970), which is not a handbook about farming and will not help you learn *how* to manage an organic homestead. But this book of poetry will surely show you *why* you should manage an organic homestead. And maybe that's just as important. Also read Berry's book of essays *The Long-Legged House* (Harcourt, Brace and World, Inc., New York, 1969) which tells you why even better. The essays in part three of the book are particularly pertinent.

Next I advise organic homesteaders to become collectors of old books on farming and gardening. By old I mean books written before 1940. Those published in the 1800s can be of particular interest since almost all farming and gardening was done organically in those days, and much know-how in avoiding bug and blight problems of that time has not been preserved by modern scientific farming. I'll list a few books I have found, which are useful, but I warn you that I know of no way to get copies except by hunting in used book stores.

T.B. Terry, *Our Farming: How We Have Made a Run-down*

Farm Bring Both Profit and Pleasure (The Farmer Company, Philadelphia, 1893). A particularly fine book for the practical organic farmer, though the author assumes his readers are already fairly knowledgeable about agriculture. Terry goes into excellent detail on subjects of: 1.) growing clover for soil enrichment; 2.) leaving residues from grain plants on the land as mulch; 3.) efficient use of manures; and 4.) bug control on large acreages without chemicals. Terry didn't use the word organic in 1893, but he was a stout defender of natural farming and opposed the use of chemical fertilizer and insecticides. And his commercial farm was very successful.

Haydn Pearson, *Fifteen Ways to Make Money in the Country* (Grosset and Dunlap, New York, 1949). ——— *Success on the Small Farm* (McGraw-Hill, New York, 1946). Pearson's two books are solid, practical "how-to-do-it" for small farmers. Remember though that the prices he quotes are very much outdated.

M.G. Kains, *Five Acres and Independence* (Greenberg Publisher, New York, 1935) Packed with information, not all of it useful today, but even that is interesting.

For rediscovering the agricultural lore and practices of prechemical days, I'm partial to *Rural Affairs,* written in nine volumes (!) by J. J. Thomas and first published in 1875 by Luther Tucker & Son, Albany, New York. I have volume four of the 1889 edition, and it is excellent on information concerning fruit, poultry, livestock, and farm buildings. The many fine drawings are also of great help.

Another quite rare set of books that contain good information is the *Biggle Farm Library* by Jacob Biggle, published by the Wilmer Atkinson Company, Philadelphia from 1895 to at least 1901. I'm not sure how many of the pocket-sized volumes are in a complete set, but there seems to be one on each kind of livestock, one on poultry (which I have), and some on various fruits and vegetables.

George D. Aiken's *Pioneering with Fruits and Berries*, (Stephen Daye Press, Brattleboro, Vermont, 1936) is not exclusively organic in scope, but is my favorite book on fruits.

For vegetables, the first part of Benjamin F. Albaugh's *Home Gardening: Vegetables and Flowers* (Grosset and Dunlap, New York, 1915) isn't bad. Neither is *Vegetable Forcing* by Ralph L. Watts, (Orange Judd Publishing Company, New York, 1917).

I can't say it too many times. Agricultural agencies of both federal and state governments have all kinds of brochures and bulletins containing information on every conceivable agricultural topic. The Extension Information divisions in the agricultural colleges of land grant universities also have information, and it is part of their job to pass it out. For bulletins from the United States Department of Agriculture, the address is: Superintendent of Documents, U.S. Government Printing Office, Washington, D.C. 20402. You can often get USDA bulletins from local Agricultural Extension Service offices—there's one in nearly every county, usually in the courthouse someplace. Look in the phone book.

Index